SOLAR POWER FOR BEGINNERS

How to Design and Install the Best Solar Power System for Your Home

Paul Holmes

Shalve Mohile

Published by Monkey Publishing
Edited by Lili Marlene Booth
Cover Design by Diogo Lando
Printed by Amazon

1st Edition, published in 2020
© 2020 by Monkey Publishing
Lerchenstrasse 111
22767 Hamburg
Germany

ISBN: 9798642013625

All rights reserved, including the right to reproduce this book or portions thereof in any form whatsoever except for brief quotations in critical reviews or articles, without the prior written permission of the publisher.

DISCLAIMER

The steps outlined in this book are based on the author's personal experience in the solar industry in the last decade. Therefore it is advised to take note of all the safety standards mentioned in this book and the safety standards in the reader's country. This book is meant for educational purposes only and helps you to learn how a solar PV rooftop system is designed and implemented. Please note that improper use of the equipment and/or procedure/s can lead to lethal damage. Thus, it is advised to take all preventive and precautionary measures to safeguard your life from all possible threats. This book is for educational purposes only, and we encourage you to seek professional advice as per specific circumstances and requirements before implementing or acting upon any information contained in this book. Further, we make no claims, promises, or guarantees about the accuracy, completeness, or adequacy of the content of this book, and disclaim liability for errors and omissions. Any action that the readers may take upon the information contained in this book shall be entirely at the readers' own risk. We shall not be liable for any damages or losses in connection with the use of this book. Each country has a different set of electrical standards and compliances' which are meant to be followed before installing solar on the rooftop. Different states have different procedures for the installation of solar. Without the approval of the state authorities, installing a solar plant is illegal. Hence, it is advised that after the designing of the plant on paper/ software based on this book, the reader shall take permission from the local authorities. Popularly known as permit package, the reader is advised to make a permit plan based on the designs in this book and submit and verify this plan with the authorities before installing solar.

Table of Contents

DISCLAIMER iii
CHAPTER 1: INTRODUCTION 9
  1.1 Solar Energy 9
  1.2 Installing rooftop solar has several benefits: 9
    1.2.1 Do I need any prior experience of engineering techniques? 10
    1.2.2 How much cost saving can I get? 10
    1.2.3 How much money can I save if I install solar panels on my own? 10
    1.2.4 How much time will it take for me to plan and install my solar panels? 10

CHAPTER 2: BASIC TERMINOLOGIES 11
  2.1 Important Terminologies 11
    2.1.1 So, what happens when you open the tap slowly? 12
    Highlights from this chapter: 12

CHAPTER 3: WHAT ARE SOLAR PANELS MADE OF? 13
  3.1 Solar Panel Components 13
    3.1.1 Frame 13
    3.1.2 Solar Cells 13
    3.1.3 Glass 14
    3.1.4 Encapsulant 14
    3.1.5 Back sheet 14
    3.1.6 Junction box 14

CHAPTER 4: HOW DOES A SOLAR PANEL GENERATE ELECTRICITY? 15
  4.1 Working of a Solar Cell 15
  4.2 How it Works (At Micro-scale) 16
  4.3 How it Works (At Macro-scale) 19
  4.4 Interconnection of Solar Panels with Electric Load 20
    4.4.1 What is the difference between AC and DC? 21
    4.4.2 What is an inverter? 21
    4.4.3 What is the difference between AC and DC? 21
  4.5 Is that it? What about solar panel size? 23

CHAPTER 5: PANEL SIZING 24
  5.1 Do I need to read everything? 24
  5.2 Page 2 is where the gold lies! 24
    5.2.1 Explanation of the rule of thumb: 26
    5.2.2 What if I am using 375W panels? 28
  5.3 Can we generalize this into a formula for quick analysis? 28

CHAPTER 6: ROOF SIZING 30
  6.1 Step 1: Understanding the equator 30
  6.2 Step 2: Measuring your roof size 34
    6.2.1 Necessary steps for calculating the area of the roof 34
      6.2.1.1 Calculate the length of the roof 34

  6.2.1.2 Calculate the width of the roof 35
  6.2.1.3 Step 1: Fire Safety Setback: 35
  6.2.1.4 Step 2: Obstructions 38

CHAPTER 7: PANEL ORIENTATIONS 41
 7.1 STEP 1: Identify the type of your roof 41
 7.2 Panel placement on a pitched roof 42
 7.3 Panel placement on a flat roof 42

CHAPTER 8: INTERROW SPACING 45

CHAPTER 9: PANEL SELECTION CRITERIA 47
 9.1 Criteria of Solar Panel 47
  9.1.1 Cost: 47
  9.1.2 Warranty: 48
  9.1.3 Efficiency/Technology: 50
  Checklist: 52

CHAPTER 10: SOLAR PANEL MAINTENANCE 54

CHAPTER 11: THE INVERTER AND ITS LOSSES 55

CHAPTER 12: INVERTER SELECTION CRITERIA 56
 12.1 Introduction 56
 12.2 What is a Phase? 56
  12.2.1 What is Single Phase? 60
  12.2.2 What is Three Phase? 61
  12.2.3 Single Phase or Three Phase 63
 12.3 Inverter sizing 64
 12.4 Hidden losses 64
  So, how do I select an inverter then? 65
  Checklist for selection of inverters: 67

CHAPTER 13: THE BATTERIES 68
 13.1 The Benefits of Using Solar Batteries 68
 13.2 Battery Characteristics 75
  13.2.1 Battery Requirements 75
 13.3 Battery Basics 75
  13.3.1 Battery Cells and Packing 76
  13.3.2 Efficiency 76
  13.3.3 Power Rating 77
  13.3.4 Battery Capacity 77
  13.3.5 Depth of Discharge (DoD) 77
  13.3.6 State of Charge (SoC) 77
  13.3.7 C-rate 78
  13.3.8 Lifecycle 78
  13.3.9 The Lifespan of Storage Devices 79
  13.3.10 Specific Power 79
  13.3.11 Energy Density 79

- 13.3.12 Power Density 80
- 13.3.13- Manufacturers and Performance Warranties 80
- 13.4 Types of Batteries 80
  - 13.4.1 Lead Acid Battery 80
  - 13.4.2 Nickel Metal Hydride (NiMH) Battery 81
  - 13.4.3 Lithium-ion (Li-ion) Battery 82
  - 13.4.5 Vanadium Redox Flow (VRB) Battery 84
  - 13.4.6 Comparative Analysis of Batteries 85

CHAPTER 14: How to Install Batteries? **88**
- 14.1.1 Series Combination 88
- 14.1.2 Parallel Combination 89
- 14.1.3 Series-Parallel Combination 90
- 14.2 Battery Sizing for PV System 92
- 14.3 Conclusion 92

CHAPTER 15: READING TECHNICAL DATASHEETS **94**
- 15.1 Reading solar panel datasheets: 94
  - 15.1.1 Nominal Power (Wp): 95
  - 15.1.2 Watt Class Sorting: 95
  - 15.1.3 Nominal power voltage (Vmp) 95
  - 15.1.4 Nominal power current (Imp) 96
- 15.2 What is this connection efficiency that we are talking about? 96
  - 15.2.1 Open circuit voltage (Voc) 96
  - 15.2.2 Short circuit current (Isc) 97
  - 15.2.3 Efficiency 97
- What is STC? 98
- What are these standard test conditions? 98
- But what if my temperature is different to that and what is irradiance? 99
- What's the point in installing solar if I am getting less output? 99
- Do I need to redesign everything? 101
- Unit of measurement: 103
- Length of the panel: 103
- Width of the panel: 103

CHAPTER 16: READING INVERTER DATASHEETS **104**
- What is MPPT? 105
- What is MPPT range? 105
  - Input data: 106
  - Maximum Permitted PV Power (kWp): 106
  - Number of MPPT: 106
  - MPPT Voltage Range: 107
  - Open Circuit Voltage (Voc): 108
  - Checklist: 110

CHAPTER 17: CONNECTION **112**
- Parallel connection: 113

Series connection: 115
Understanding short circuit and fire hazard 118
    What would happen if the pressure increased suddenly? What if someone forces water through the pipe downwards towards the outlet forcefully using something like a motor? 118
    Which tap will burst first? 119
    Which conductor will current flow through? 119
    At this point you might be wondering, why does a conductor catch fire in a short circuit condition? 120
    What will happen if I use a wire which has a capacity of 18A? 120

CHAPTER 18: PROTECTION DEVICES **122**
  MCB: 122
    "Which MCB should I choose?" 123
  Fuse: 124
    Isolator: 125
    Case 1: Repair/ maintenance needed at the Solar Panel Array 126
    Case 2: Repairs needed at the inverter 126
    Case 3: Repair/ maintenance needed at the Main Service Panel ... 126

CHAPTER 19: CONNECTING STRING TO INVERTER **128**

CHAPTER 20: UNDERSTANDING THE MAIN SERVICE PANEL **130**

CHAPTER 21: CONNECTING INVERTER TO MSP **134**
  Never cross connect the terminals 136
  Unidirectional: 138
  Bi-directional: 138
  MCB sizing: 138

CHAPTER 22: CABLE SIZING **140**
  Steps to use for cable sizing: 140

CHAPTER 23: RACKING IT UP **143**
  1. Portrait 143
  2. Landscape 144
  Rails: 145
  Rafters: 145
  Standoff: 146
  Mid clamp: 147
  End clamp: 147
    1. Type of roof: 148
    2. System size: 149
    3. Offline: 149
    4. Online: 149
    Steps to draw the rough diagram in a graphics package 149

CHAPTER 24: RACKING WEBSITES **153**

CHAPTER 25: THE FINAL CHECKLIST **158**
   Step 1: Calculate the size of the system required 158
   Step 2: Roof assessment 159
   Step 3: Selection of solar panel brand 160
   Step 4: Selection of inverter 160
   Step 5: Check for MSP upgrade 161
   Step 6: Circuit breaker selection 161
   Step 7: Cable sizing 161
   Step 8: Estimate the mechanical requirements 162
   Step 9: Buy the items 162

CHAPTER 26: INSTALLATION **163**
   Mechanical mounting 163
   Electrical connection 164
      AC MCB Sizing 167
      "But how do I know if my solar array is generating power?" 169
      Cable sizing: 169

CHAPTER 27: ADDITIONAL FACTS **171**

CHAPTER 28: ARE WE DONE? **172**

ABOUT THE AUTHORS **173**

CHAPTER 1: INTRODUCTION

1.1 Solar Energy

Going **solar** has major financial benefits: it reduces your monthly electricity costs and can even increase the value of your **home**. An incentive such as the federal tax credit for **solar** can reduce your net cost by 26 percent or more, but **solar** is still a big investment, and the price tag can result in sticker shock. When calculating the total price, consider how much energy you regularly consume — your usage is listed on your monthly utility bill — and what size system will generate the amount needed.

The rising cost of electricity from traditional sources still makes solar installation a no-brainer for many homeowners. With the help of this guide, you will learn what to consider before going solar and what steps will be involved. It will guide you along the process to install your own solar panels at home. So you can start living a more self-sufficient life.

This book will provide you with a step-by-step blueprint for designing your own rooftop solar system. I have dedicated individual chapters for each step in the installation process. Onsite images, pie charts and flow charts are presented so that the understanding of the installation process becomes easier.

Get ready to embark on a journey that takes you through the various processes and design methodologies with simple and easy to understand terms.

Since I want to make it as easy as it is to learn a new language as a child, we keep the tone of this book rather informal and easy to understand. So even someone without any background in solar can get started and most importantly have fun learning!

1.2 Installing rooftop solar has several benefits:

- Cost savings

- Closer to energy independence
- Environmentally friendly
- Access to remote locations
- Carbon savings
- Free from operational fuels

1.2.1 <u>Do I need any prior experience of engineering techniques?</u>

- *No, it will help but is not necessary*

1.2.2 <u>How much cost saving can I get?</u>

- *70%-80% of your current electricity bill (depending on your solar installation set-up)*

1.2.3 <u>How much money can I save if I install solar panels on my own?</u>

- *Anywhere from 20% to 25% of the project value*

1.2.4 <u>How much time will it take for me to plan and install my solar panels?</u>

- *It usually takes two to three days for small scale systems, but can take from a week to a month for a large scale system. It depends upon the different factors such as...*
  - *The transportation of equipment and devices from site area or installation place*
  - *The time duration required for studies such as a technical study, financial study and environmental study*
  - *Time based on scale which means that a small system will take less time to install and test compared to a large system*

CHAPTER 2:
BASIC TERMINOLOGIES

2.1 Important Terminologies

Before we get into detail on the installation process it's helpful to have an overview of the basics of electricity.

Voltage: This is the electric potential of the circuit. It's the potential energy of the circuit.

Current: This is the flow of electrons. It can be seen as the kinetic energy of the circuit. Current is the element that actually flows through a circuit.

Resistance: This is the opposition to the flow of current. It can be compared to friction while driving a car.

Let us understand this logic with an example:

Imagine a tall building with 9 floors. There is a water tank on the roof of the building. People living on all the floors get water through this tank. As soon as you open the tap, the water starts flowing through the pipe connected from the tank to your pipe. Hence, you get water as soon as you open the tap. If the pipes are clear, who among the following will get the water at their tap at maximum intensity?

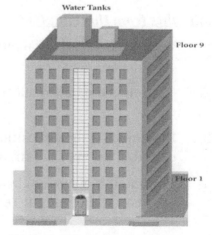

Figure 2.1 Illustration of terms

1. Floor 1.
2. Floor 9.

If you observe closely, the people living on floor 1 will get the water at maximum intensity as there is a huge gravitational "potential difference" between the two. The gravitational potential difference between Floor 9 and the tank is less compared to that between Floor 1 and the water tank.

Similarly, in an electric circuit the potential difference between any two points is known as voltage. The current can be the flowing water. We can infer that the greater the potential difference between the two points, the higher the intensity of the flow will be. Similarly, the higher the voltage between any two points, the higher the current will be given the same resistance (more on this later in the chapter on design).

2.1.1 So, what happens when you open the tap slowly?

The speed at which the water flows through the tap varies per the tap opening. The valve decides the opposition ("resistance") which will be placed against the flowing water. Similarly, in an electrical circuit, the amount of current is decided by the resistance of the circuit.

Highlights from this chapter:

1. Voltage is the potential difference/ potential energy of the circuit.
2. Current is the flow of electrons/kinetic energy of the circuit.
3. Resistance is the opposition to the flow of current in the circuit/friction of the circuit.
4. The amount of current in a circuit is dependent on the applied voltage and the resistance offered to it.

CHAPTER 3:
WHAT ARE SOLAR PANELS MADE OF?

3.1 Solar Panel Components

I would like a cheeseburger with two extra layers of cheese!

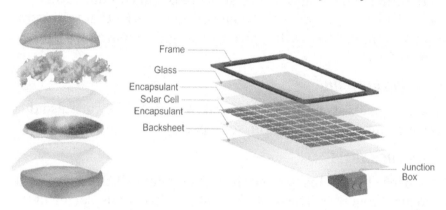

Figure 3.1 Illustration of a Solar Panel

Just like your multi-layered sandwich, solar panels are also made up of many layers, each of which has a unique role to play.

3.1.1 Frame

The frame helps to hold all the other layers of the solar panel together. It also helps in preventing any damage that might occur during the transportation of the panels. The most common materials used for the solar panel are either power coated aluminum or anodized aluminum.

3.1.2 Solar Cells

The solar cells are made up of semiconductors. Different types of materials are used such as silicon (Si), germanium (Ge), cadmium sulphide (Cds), gallium arsenide (GaAs) with the most commonly used material being silicon. The solar panels work on the principle of photovoltaic effect. The semiconductors are capable of converting the light energy of the sun into electrical energy. This is where the actual conversion of energy takes place.

3.1.3 Glass

The glass is the first line of defense. This protects the solar cells from dust, dirt and other pollutants. The glass is made of anti-reflective material. This anti-reflective material prevents the rays of the sun from being reflected so there is a maximum absorption of the sun's rays by the solar panels. The material that is usually used is tempered glass. It is stronger when compared with other types of glass.

3.1.4 Encapsulant

The encapsulants hold the solar cells together. They are also the second line of defense for the solar panels.

3.1.5 Back sheet

The back sheet helps to keep the frame, solar cells, encapsulants and glass together. The back sheet also helps in the placing of the junction box.

3.1.6 Junction box

This is the output of the entire solar panel. The most commonly used material for the cable is copper. One end of the copper wire is connected to the solar cells internally while the other end of the copper wire is taken as the output. The diode is also provided as protection.

CHAPTER 4:
HOW DOES A SOLAR PANEL GENERATE ELECTRICITY?

Sunlight contains photons. Photons can be seen as tiny packets of energy. These tiny packets of energy impact on earth every day.

Just like chlorophyll pigment in the leaves of plants absorbs sunlight and converts it into energy, our solar panels absorb photons and convert them into electrical energy. The illustration is provided in the figure 4.1:

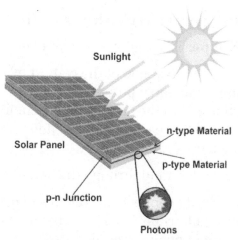

4.1 Working of a Solar Cell

Figure 4.1 Solar panel's operation

From our chapter on basic terminologies, we have realized that we need to have a potential difference between any two points for the current to flow. So whenever you see a battery, there are two terminals on the battery. These two terminals are at different potentials. If a battery is marked 12 volts, this means that the potential difference between them is 12 volts.

Solar cells are just like a battery. They provide us with two terminals which are a certain voltage or which are at two different potentials. But this battery works only in sunlight. Before we dive deeper into the workings of the solar cells, let us make a few things clear,

1. We need a potential difference.
2. If we have a potential difference, it will lead to the flow of current (flow of electrons).

3. The flow of current through any device is a flow of power/energy.

4.2 How it Works (At Micro-scale)

Remember, our sole job is to create a potential difference and make electrons flow (current flow) through our appliances.

So, how do we generate this potential difference in a solar cell?

Solar cells are made up of two types of material regions.

P- type: This region has holes (Positive Charge).

N-type: This region has a lot of electrons (Negative Charge)

Just like our example of the water tank, if there are two different regions of unequal potentials, they will try to reach an equilibrium. In the tank example, equilibrium was when the water reached each respective floor and finally the ground through the drain.

Let us understand through the below analogy.

Similarly, if I connect two regions of different potential, I will reach equilibrium. Let's see what happens when we connect two regions of unequal energies.

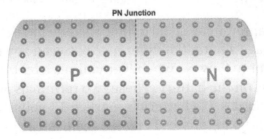

Figure 4.2 PN junction diode

These are two different types of materials. P-Type and N-Type.

- **N-Type**: Has more electrons (Negative Charge)
- **P-Type**: Has fewer electrons (Positive Charge)

What will happen if I connect them together?

Do they reach an equilibrium?

For them to reach equilibrium there has to be some energy to it. In our tank example, it was gravity. Similarly, here it is an electronic force of attraction.

So, let us connect them and see what happens.

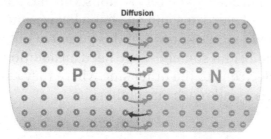

Figure 4.3 Diffusion in PN junction diode

All atoms have electrons. Some have extra electrons while some are deficient in electrons (holes). Therefore, when they come into contact with other materials, they try to reach a state of equilibrium. Similarly, when an N-Type material that has more electrons comes into contact with a P-type material that has fewer electrons (holes), they share their electrons in such a manner so as to reach equilibrium.

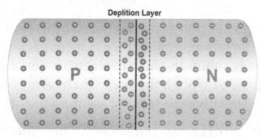

Figure 4.4 Formation of depletion layer in PN junction diode

We have reached equilibrium. Both the materials (P-Type & N-Type) have equal energies. As you can see, there is a neutral layer in between (this is called a depletion region). Now the positive charges from the p-side cannot reach the negative charges on the n-side due to this depletion region.

But there is a problem. Now the electrons from the n-type region cannot travel internally to the p-type region since there are already electrons at the center of the p-n junction and so there is a potential difference in the two regions. One has a higher concentration of electrons and one has a lower concentration of electrons. Now that we know that this potential difference has to reach an equilibrium somehow, we need an additional path for the flow of electrons to the other side where they can meet the positive charges and reach an equilibrium.

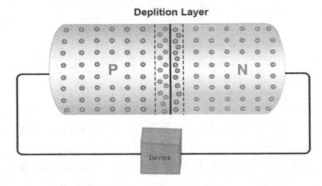

Figure 4.5 Behaviour of PN junction diode in connection with device

Remember, opposite charges attract each other, but they cannot overcome the depletion region. If we just connected a wire as above, it would result in the flow of electrons (energy) from the n-side to the p-side through the wire. These electrons can then meet the holes (positive charges) present on the other side.

Device: TV, AC, LIGHTS, etc. (any household appliance working on electricity).

There is a flow of energy through the above circuit but there is one problem we need to solve. The energy flowing through the circuit is very small and once the electrons travel towards the positive side there will be no electrons left! In this case, we need a source that can keep on generating more and more electrons;

in our case that source is sunlight. Considering a real-life example this is like giving a bodybuilder only one egg to eat as the energy to do his workout, but once he consumes the energy from that egg he would need a constant supply of energy for his next workout!

4.3 How it Works (At Macro-scale)

To summarize, just as the bodybuilder needs a constant supply of energy to keep working out, a solar cell would need a constant supply of electrons to maintain the potential difference and keep supplying our load (appliance). This force or energy to keep the electrons going is sunlight. When sunlight is present on the surface of the material, it acts on the depletion region to create more electrons that move through the load.

To summarize what we just found out:

1. We created potential differences using two materials.
2. The potential difference itself is not enough because we need a force to make the electrons move to the other side where they can meet the holes. This force is sunlight.

Let us understand step-by-step what exactly happens when the light strikes the solar panels.

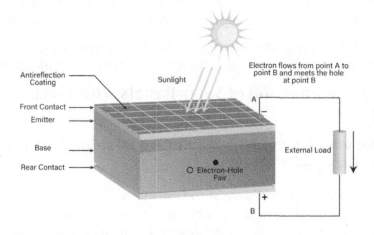

Figure 4.6 Conversion of Solar Energy to Electricity

The solar panel is made up of solar cells. When the rays of the sun (referred to as solar irradiance in technical terms) strike the solar cells, the electrons become energized and try to reach the holes. Since there is a junction (blockage) already in between, the electrons need an external path to flow. Hence, the electrons flow through the external path.

This external path is where we connect our appliances (also known as the load). The flow of electrons through our appliance will make it work. Hence, we connect our appliance in between this path.

Many such small cells combine to form a solar panel. Now we know one thing for sure, that we need to connect our appliances in-between points A and B, so that the energy can flow through our appliance.

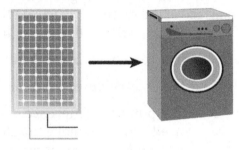

These points A and B are nothing but the two wires coming out of your solar panel. This is the output of the solar panel where we get the required energy to power our equipment.

Figure 4.7 PV and device connections

The output contains two terminals. As you can see in the image below, there is a red-colored wire and the other is the black wire.

4.4 Interconnection of Solar Panels with Electric Load

So, can we just connect the output of solar panels to our appliances?

Though the energy available at the output terminal is useful, it is not in the desired format. The output power is DC power, whereas what we require is AC power. Let us understand that with the help of an example.

20

4.4.1 What is the difference between AC and DC?

- Imagine there are two people working for the UN. One is from Japan and the other one is from the USA. The former can speak in Japanese only and the latter can speak in English only and not vice-versa. Both are required as not everyone knows each language
- (Similar to our above example, AC and DC can be seen as two different forms of power in electricity. While solar panels generate DC (speaks Japanese) the home appliances require AC (speaks English)

4.4.2 What is an inverter?

- What would happen if the Japanese person were to communicate with the American? Neither can speak the other's language, right?
- As you must have guessed, there is a need for a translator in-between. The translator shall translate Japanese into English
- (Continuing with our example, the inverter is a device that translates or converts the DC power into AC power). It helps in converting the power generated by the solar panels (DC power) into a format that can be accepted by home appliances (AC power)

4.4.3 What is the difference between AC and DC?

DC: In direct current (DC), the electric charge (current) only flows in one direction.

AC: In alternating current (AC), the electric charge changes direction periodically.

- Most of the household appliances that we use such as televisions, microwaves, fans and lights require AC power at the input. So we need a device that converts the direct current from PV Modules to Alternating Current. That device is known as an inverter.

Inverter: The inverter is a complex electronic/ electrical device which takes in DC power from the modules and converts it to usable AC power for appliances.

The power available at the output cannot be used directly. So now that we know that we need an inverter, what would our block diagram look like?

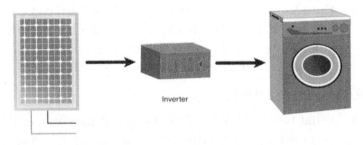

Figure 4.8 Interconnection of PV panels and electric load through inverter

☼ Inverters Explained

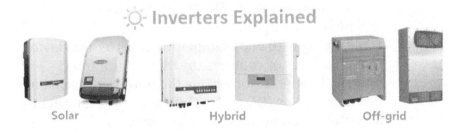

Figure 4.9 Inverter types

The three main categories of inverter are:

a. Solar Inverter: This is directly connected with PV panels at input and to the load at output
b. Hybrid Inverter: This is an advanced technology which stores excess energy in the battery, and is connected to the electricity grid
c. Off-grid: This allows the system to rely on itself and has no connection to the electricity grid

4.5 Is that it? What about solar panel size?

Let's assume you want to power your washing machine using solar. The power generated by a solar panel varies depending on size. Let's take a random solar panel. Trina 250 Watt (W).

Trina: Brand name (can be different depending on location and availability in your area).

250W: This is the amount of power that the solar panel is capable of generating.

W = watts = Unit of power.

This means that the above-mentioned solar panel will generate 250W of power. The power required by your typical washing machine is 500W. If I were to ignore the losses for a moment, what we have is a situation something like this:

Generated power by solar panel= 250W.

Required power by washing machine= 500W.

Therefore I would need two solar panels each of 250W.

There you have it. You have roughly estimated the capacity required to power your washing machine. Now you must be wondering, I have so many appliances TV, AC, heater, etc. and not just the washing machine, so how do I calculate my consumption? How will I know how many panels will be required to meet my energy needs? This is where we begin our first step in the design process called, *"SIZING."*

CHAPTER 5: PANEL SIZING

In the previous chapter, we roughly estimated the number of solar panels required to power our washing machine. By the end of this chapter, you will know the exact number of panels required to power all the appliances of your house. In order to do that we need to know two things just as we did in our previous example:

- Energy requirement
- Energy generation

Let us calculate the energy requirement first. The following items are required:

- Electricity bill
- Calculator
- Pen and paper

5.1 Do I need to read everything?

No, you don't. If you look closely there will be two pages to every bill. Page 1 will mention the following things:

- Amount to be paid
- Address
- Discount and rebates
- Payment method
- Billing date

All of the above are generally mentioned on page one. This page is not useful for us in determining the capacity required.

5.2 Page 2 is where the gold lies!

Let's take a closer look at page 2 and try to understand it.

Take a look at the charges section of page 2 and just focus on the following,

1. Total consumption in kWh.

2. Bill period.

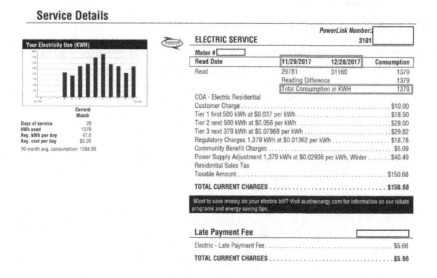

Figure 5.1 Electricity Bill

Write down the two figures and keep the bill inside your pocket as these two data points are enough to calculate the number of panels required.

In the given bill the following can be observed:

- Total Consumption in kWh = 1,379 kWh = 1,379 units (1kWh= 1 unit)
- Bill Period: 11/29/2017 to 12/28/2017

Count the number of days in between the given bill period.

In this case, the total number of days= 33 days.

Alright then, did you keep the bill in your pocket?

The bill tells us that we have consumed 1,379 kWh (units) in 33 days.

Hence, let us calculate the consumption per day:

=1,379/33

= 41 units

That means we require **41 units** per day!

Now that we know the power required per day let us evaluate how many panels are required to do the job.

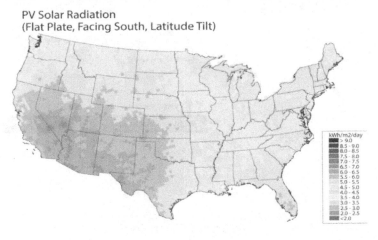

Figure 5.2 Solar Potential of the location in kWh/m2/day

Let's say we use the same brand of panels i.e. Trina 300W.

Remember the rule of thumb:

One watt produces 0.0045 (Or 1 kW- 4.5 units per day) units per day (varies per location)

5.2.1 Explanation of the rule of thumb:

Search the Solar Irradiance map of your country and locate your region. This data is easily available on the Internet via government websites.

Let us take Michigan as an example:

The above graphic tells us that on a given day, on a 1 m² area, we will get 4 to 4.5 units (At latitude tilt, facing south)

1 meter= 3.28 feet.

1 sq. meter= 10.76 sq. feet.

The dimension of a Trina 300W solar panel is 1.6 square meters. (17.22 sq. ft.).

= 4.5 x 1.6

= 7.2 units/day.

Ideally one solar panel should be capable of generating 7.2 units/day.

But the solar panel efficiency is 18%. This means solar panels can only capture 18% of the energy impacting on them.

So, one solar panel of 1.6 sq. meters area will generate

4.5 units x 1.6 sq. meter area (17.22 sq. ft.) x 0.18

= 1.30 units/day for 300W.

This means that if we place one solar panel of capacity 300W we will get 1.44 units of electricity per day.

Getting back to the rule of thumb, we now know that 300W produces 1.30 units.

Hence, one watt will produce,

1.30/300 = 0.0044 units per watt.

Hence the rule of thumb.

We don't need to work this out every time. You can directly divide the irradiance by 1,000 to obtain your rule of thumb.

And so for Michigan I will consider,

1 watt produces 0.0045 units per day (Since it says 4 to 4.5 kWh in the given region).

I have simply divided the figure by 1,000 to obtain my rule of thumb,

1 watt= 0.0045 units/day

So 300W will produce= 300*0.0045 units

= 1.35 units/day.

Therefore, if we place one Trina 300W panel we will generate 1.35 units/day.

But we require 41 units.

So we will need,

41/1.35

= 30 panels.

There you have it! You require 30 panels of 300W each.

5.2.2 What if I am using 375W panels?

A 375W panel will produce;

375*0.0045

= 1.6875 units per day.

So, if we place one 375W panel it will generate 1.6875 units/day.

But we require 41 units.

So we will need,

41/1.875

= 24 panels.

5.3 Can we generalize this into a formula for quick analysis?

$$\text{Number of Panels} = \frac{\text{Number of peak units required per day}}{(\text{Wattage} \times 0.0045)}$$

$$\text{Number of Panels (For 300W)} = \frac{41}{(300 \times 0.0045)} = 30 \text{ Panels}$$

$$\text{Number of Panels (For 375W)} = \frac{41}{(375 \times 0.0045)} = 24 \text{ Panels}$$

Congratulations, you have successfully designed the panel sizing!

As you can see, the number of panels varies depending on the wattage of the panels. For example, if your roof cannot accommodate 30 panels but it can accommodate 24 panels, then it is clear that a wise choice would be to choose 375W panels. But we need a fixed quantity to define our system size. We simply say,

I need 30 panels of 300W each.

    OR

I need 24 panels of 375W each.

    OR

I need 26 panels of 350W each.

We need a standard notation to be clear as to what exactly we need. So we must define the **SYSTEM SIZE**.

SYSTEM SIZE= Number of panels x Wattage of panels

- For 300W panels - 30 x 300= 9 kW~ 9 kW (1000W=1kW)
- For 350W panels-26 x 350= 9.1 kW~ 9 kW (1000W=1kW)
- For 375W panels - 24 x 375= 9kW~ 9 kW (1000W=1kW)

This means that our requirement is **9 kW.**

The wattage of the panel depends on many factors which will be dealt with in the chapter on **Selection of Panels**. So, from now on, we will refer to the system size as 9 kW.

CHAPTER 6: ROOF SIZING

In our previous example, we came to know that we would need 9 kW. But, is there enough space to fit 9 kW of panels on the roof? Just like we check if the clothing matches our body size we also need to check if the desired number of solar panels can fit on our roof!

We will need the following things in order to determine this...

- An understanding of the equator
- A measuring tape
- Solar panel dimensions
- A calculator

6.1 Step 1: Understanding the equator.

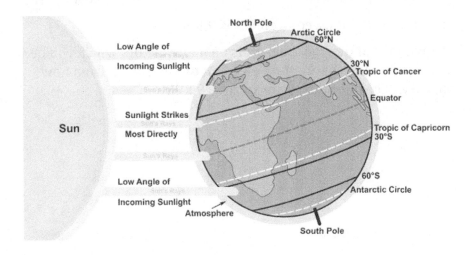

Figure 6.2 Equator and Solar Irradiance

We all know that the earth rotates around the sun.

The sun rises in the east and sets in the west. If you were to read the last statement we denote the movement of the sun and not the earth. Similarly, the equator is the imaginary line which divides the Earth at its center.

How does that affect the panel placements?

Just imagine the following two countries...

1. USA.
2. Australia.

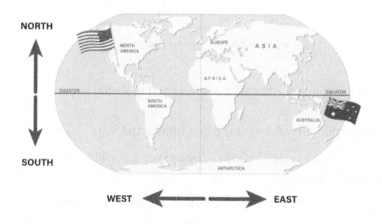

Figure 6.3 Cardinal directions

The above diagram depicts the position of both countries with respect to the equator. Let us place four people, two in each country. Let the two people in the USA be U1 and U2, while those in Australia be A1 and A2.

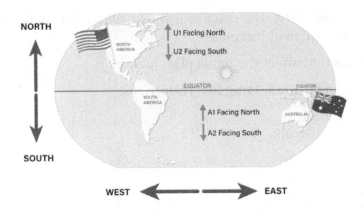

Figure 6.4 Cardinal directions with respect to standing

If you observe closely, A1 and U2 are facing the sun.

A2 and U1 are not facing the sun. These two people have their backs facing the sun.

Now imagine the same concept and just replace the people with panels.

Let there be 4 solar panels A1, A2 and U1, U2.

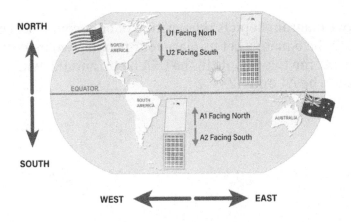

Figure 6.5 Cardinal directions with respect to PV Panels

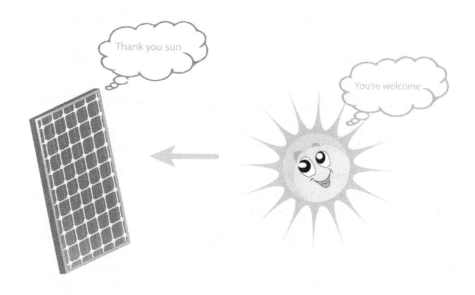

Figure 6.6 Appropriate direction of Solar Panel

A1 and U2 happily generating power.

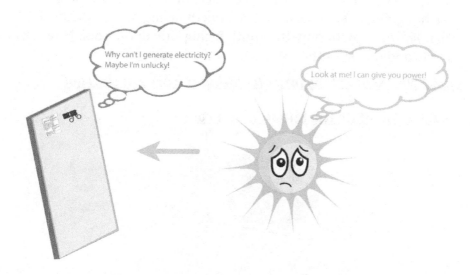

Figure 6.7 Inappropriate direction of Solar Panel

A2 and U1 not able to generate power.

If panels A2 and U1 are not facing the sun they will not absorb the power of the sun. Hence there is no point in placing these panels. we can conclude that the panels should never be facing towards the opposite side of the equator.

Never place panels facing opposite to the equator!

Hence, if you live in a country above the equator you can place the panels on the following roof faces:

1. South (First preference).
2. West (Second preference).
3. East (Third preference).

And, if you live in a country below the equator you can place the panels on the following roof faces,

1. North (First preference).
2. West (Second preference).
3. East (Third preference).

6.2 Step 2: Measuring your roof size

Imagine you have a roof like this. Since, this is a roof in the US, we are not going to spend time on the north facing part of the roof. That leaves us with only the south facing side of the roof. Now, go ahead and measure the total area of the roof.

6.2.1 Necessary steps for calculating the area of the roof

6.2.1.1 Calculate the length of the roof

Figure 6.8 Calculation of roof length

6.2.1.2 Calculate the width of the roof

Based on the above roof measurement we know that the area of the roof = 40.8 x 18.4 = 750.72 sq. ft. (69.74 sq. m.)

But there is a catch! We cannot utilize the roof area fully as we need to consider the below factors...

Figure 6.9 Calculation of roof width

1. Fire safety setback.
2. Area of roof obstructions.

6.2.1.3 Step 1: Fire Safety Setback:

What is fire safety setback?

In case of any fire incident on the roof there has to be enough space on the roof for the firefighter to take hold of the situation and extinguish the fire. The fire safety setback in various countries differs and readers must check with local authorities for this information. For instance, in many states in the US the setback is 1.2 meters (4 ft.).

Figure 6.10 Fire safety setback (width)

To mark the setbacks, let us begin with the bottom edge of the roof. The bottom edge of the roof will need no setback as it can be directly accessed from the front. But the other sides do need a setback. Consider the east and west sides of the roof. Each will require a setback of 1.2 meters (4 ft.). The panels cannot be

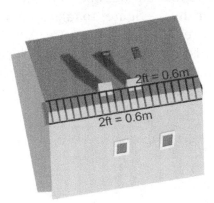

Figure 6.11 Fire safety setback (length)

placed in the shaded regions shown below.

Now take a close look at the center of the roof. In this case we need a 0.6 meter (2 ft.) setback from each side of the roof i.e. 0.6

meter setback from the north facing side and 0.6 meter setback from the south facing side. Here is what this would look like: Finally, our roof will look like this considering all the setbacks.

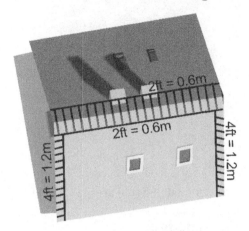

Figure 6.12 Consideration of safety setbacks

Now let us calculate the length and width excluding the setbacks.

The total roof area without excluding the setback is 750.72 sq ft (69.74 sq m).

The total roof area excluding the setback would be equal to total roof area- total area of setbacks.

Total area of setbacks = area of setback at left side + area of setback at right side + area of setback at the top.

Total area of setbacks = roof width x setback (left) + roof width x setback (right) + roof length x setback (top).

Total area of setback = 18.4 ft x 4ft +18.4ft x 4ft+ 32.8 ft x 2 ft

= 212.8 sq ft

So, the total roof area = 750.72 -212.8 = 537.92 sq ft (50 sq m)

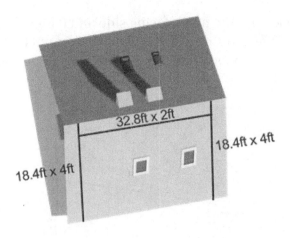

Figure 6.13 Remaining roof area after safety setbacks

6.2.1.4 Step 2: Obstructions

If you observe closely, there are obstructions on the roof.

Figure 6.14 Consideration of Obstructions

There can be skylights, chimneys and other ducts, or vents which prevent us from placing the solar panels on the roof. So our useful area is not the full 50 sq m (537.92 sq ft). Our useful area now will be:

50 - Area of obstructions.

How to calculate the area of obstructions

You can estimate the area of obstructions in a similar way to how we have estimated the roof area. Consider the obstructions to be the same as the roof i.e. rectangular in shape.

On our roof, let's assume the obstruction to be= 2.5 sq m (26.90 sq ft).

There are two such obstructions on this roof. Therefore, the total area to be subtracted will be 2.5 x 2 sq m.

Now the useful area is 50-5= 45 sq m (484 sq ft).

Finally, we are left with a 45 sq m (484 sq ft) area in which we can place our solar panels.

RULE OF THUMB:

1 kW requires a 6-8 sq m (64-86 sq ft) area.

In the previous chapter we came to know that we need a system size of 9 kW. So the area required will be...

    8 x 9 (Consider 8 sq m per kW to be on the safe side)

    = 72 sq m. (775 sq ft).

    Now there can be two possible cases:

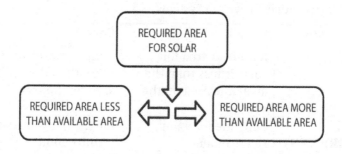

Figure 6.15 Requirement of Solar Area

Case 1: Required area is less than the available area.

Case 2: Required area is more than the available area.

Let us evaluate each case individually.

Case 1: Required area is less than the available area

Depending on the type of roof, it may happen that the required area for solar is less than the available area on your roof. So we can successfully offset the majority of our electricity bill resulting in maximum savings.

Case 2: Required area is more than the available area

Sometimes, the roof size is not enough to meet the required area demand. In this case we need to place the maximum number of panels possible. Let's see how many panels we can fit...

As per the rule of thumb, each kW of solar requires an area of 8 sq m (86 sq ft) and we have 45 sq m (484 sq ft) of roof space. So the maximum amount of solar we can fit on our rooftop will be:

$$45/8 = 5.6 \text{ kW}$$

Hence, we can still fit a little above 5 kW of solar on our roof!

Our requirement is 9 kW and we can manage to fit around 5 kW. The following are the consequences of installing part of the required capacity vs. required capacity.

Table 1: Comparative Analysis of Systems

Comparison	Case 1	Case 2
AREA	Required area for solar is less than the available area on the roof	Required area for solar is more than the available area on the roof
SIZING REQUIREMENT	Fully meets sizing requirement	Partly meets the sizing requirement
SAVINGS	Maximum savings possible	Partial savings possible
SYSTEM COST	Comparatively high	Comparatively low

CHAPTER 7: PANEL ORIENTATIONS

Before you go on your roof to place the panels, let us first understand in how many different ways we can actually place the panels.

7.1 STEP 1: Identify the type of your roof

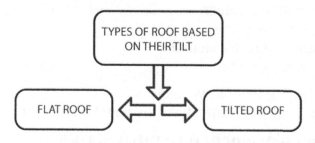

Figure 7.1 Roof types

Here is what a flat roof looks like.

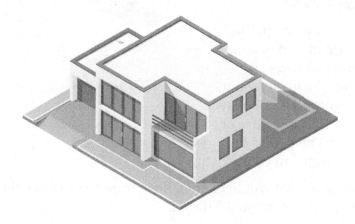

Figure 7.2 Flat roof type

As you must have noticed, the roof has no inclination whatsoever. That is why it is known as a flat roof. Flat roofs are easier to climb onto and work with compared to a tilted roof.

The above image is of a tilted roof. It is also known as a pitched roof. If any roof has a slight inclination it comes under this category. It is slightly more difficult to mount panels on this type of roof.

Figure 7.3 Tilted roof type

In the following chapter we take a look at how the panel tilt affects the panel placement.

7.2 Panel placement on a pitched roof

The arrangement made on a pitched roof is known as **a FLUSH MOUNT ARRANGEMENT.**

In this arrangement, the panels are placed next to each other in rows and columns. The flush mount system forms a second layer of tiles on the structure of your roof. The panels fit exactly parallel to your existing roof.

Figure 7.4 Flush mount arrangement

Just imagine what would happen if we tried to place the flush mount arrangement on a flat roof.

7.3 Panel placement on a flat roof

On a flat roof we cannot place the panels in a flush mount arrangement as the panels will be facing upwards towards the sky. As we have learned in the chapter on the understanding of horizon, we need to place panels facing towards the sun and so an array of solar panels facing towards the sky is not an ideal arrangement to go with. At the most, this arrangement might give us output at 12 noon when the sun is directly overhead.

Do for a flat roof, we design a different type of arrangement known as a fixed–tilt arrangement which looks like this:

Figure 7.5 Fixed-tilt arrangement

As you can see in the image above, the panels are placed on a tilt, at an angle facing skywards. The tilt for residential projects for a flat roof is generally kept at 10 degrees. It will look just like the image above.

Here's a question for you:

What will happen if we place the second row of panels just after the first row, leaving no space in between?

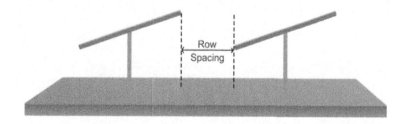

Figure 7.6 Row spacing among solar panels

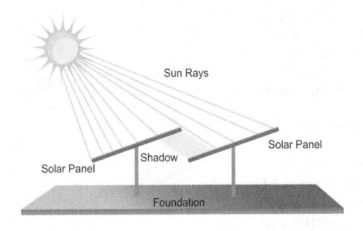

Figure 7.7 Inappropriate and shadow spacing

As you can see in the image above, the shadow of the first row falls on the panels of the second row. Hence, the area under shadow cannot produce power, so we need to keep space between two rows. This space is what we call ***INTERROW SPACING.***

CHAPTER 8: INTERROW SPACING

RULE OF THUMB:

The distance of interrow spacing= 2 x height of the panel.

Let us understand how to evaluate this. The height of the panel refers to the height from the lowest point of the panel and not the total height.

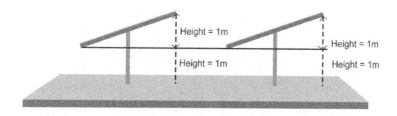

Figure 8.1 Row spacing among solar panels

For example, in the above scenario, if the highest point of the panels is 1m above ground + 1m from the lowest point of the panel then the panel height is considered to be 1m and not 2m.

Let's say that after installing the panels you calculated the height from the base of the panels to the top of the panel to be 1m. Using our above rule of thumb;

The interrow spacing

= 2 x 1

= 2 m (6.56 ft.).

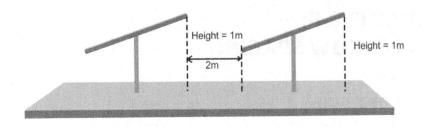

Figure 8.2 Interrow spacing

Our placement of modules will look like this:

Therefore, the interrow spacing between the two rows will be 2 m (6.56 ft).

CHAPTER 9: PANEL SELECTION CRITERIA

9.1 Criteria of Solar Panel

Solar panels come in a variety of wattages, sizes, and types. It is important to choose the one which best fits your needs. In this chapter we evaluate which solar panel is best suited for your requirements. Here is what you need to do:

Evaluate your solar panel based on these 3 parameters:

1. Cost.
2. Warranty.
3. Efficiency/Technology.

The market is filled with solar panel manufacturers. Some of them are listed below:

- REC
- Hanwha Q-Cells
- Trina
- Sun Tech
- Jinko

Now we have to compare the second list based on the parameters given in the first list. The five solar panel manufacturers mentioned above are the most widely used and reputable ones. We have to compare each manufacturer based on cost, warranty, and efficiency.

At this point in time, you must be wondering why we have selected these three parameters only, and what the significance of these three parameters is? Let us understand them individually...

9.1.1 Cost:

60% of our system cost will be from solar panels. If you choose to use batteries for your system that will add more cost. But as not everybody uses batteries for their system we ignore those cost for now and talk in more detail about it in the chapter on batteries.

It is important to select a manufacturer that gives us an optimum cost without sacrificing the quality of the system.

9.1.2 Warranty:

There are two types of warranty: I

- **Workmanship warranty:** This is the same warranty that you get when you buy a new phone. In case of mechanical deterioration of the solar panel e.g. unintentional breakage of the panel glass or frame, the manufacturers will repair that for you.
- **Warranty period:** 10 Years (Product Warranty) and 25 Years (Linear Power Warranty).
- **Warranty criteria:** The mechanical damage caused must be natural, i.e. panels damaged during installation or transport do not qualify for this type of warranty.
  The panels are generally rated for wind speeds up to 200 km/hr in case of cyclone/ natural calamity resulting in the damage of the solar panels due to higher wind speeds and/or other reasons, this warranty cannot be claimed.
- **Production guarantee:** 'Most panel manufacturers give a power production guarantee of 25 years. Let's say you buy a 300W module. This guarantee states that for the panel given above it shall produce 300W of power every year with up to 0.7% deterioration per year. The panel manufacturer basically assures us that we shall get the following output every year. Let's say we install solar panels in the year 2020. We shall get the following as output.

Table 2 Production Output in different Years

Year Number	YEAR	OUTPUT
Year-1	2020	300 W
Year-2	2021	298 W
Year-3	2022	296 W

Year-4	2023	294 W
Year-5	2024	291 W
Year-6	2025	289 W
Year-7	2026	287 W
Year-8	2027	285 W
Year-9	2028	283 W
Year-10	2029	281 W

It is to be noted that the solar panel's performance reduction depends upon different factors such as temperature, the way it is cleaned and maintained, and its deterioration. So, the results presented in Table 2 can vary from region to region. There is just a slight change in the above calculation. There is something called LID.

LID Stands for Light Induced Degradation.

When we go into sunlight after sitting inside a dark room for a couple of hours, we experience intense sunlight. This causes pain in our head as our body is not yet ready to come out of the darkness of the room. Even solar panels experience the same when they come into sunlight after they have spent a lot of time in the factory. The panels degrade due to the sunlight.

But wait, don't solar panels love sunlight?

As funny as it may sound, when solar panels come into contact with the sunlight for the first time, they lose up to 2.5% of their efficiency. But don't worry, this happens just once. Only during the first year do we experience 2.5% LID loss. After that, just 0.7% every year. Taking this into account, our new table will look like this:

First year losses= 2.5% LID losses + 0.7% losses.

Second year onward= 0.7% losses.

Table 3 Production Output after consideration of Losses

Year Number	YEAR	OUTPUT
Year-1	2020	290 W
Year-2	2021	288 W
Year-3	2022	286 W

Year-4	2023	284 W
Year-5	2024	282 W
Year-6	2025	280 W
Year-7	2026	278 W
Year-8	2027	276 W
Year-9	2028	274 W
Year-10	2029	272 W

As you can see, the panel output decreases with time, but the panel manufacturer guarantees us the above output. The power production guarantee is generally valid for 10 years which is why the table above covers 10 years. The above guarantee is valid at STC conditions.

STC conditions: 25 °C cell temperature, irradiance 1000 W/ sq m (more on this later).

9.1.3 Efficiency/Technology:

The efficiency of solar panels refers to how efficiently they can convert the sunlight into electrical energy. Let us understand this with an example.

Suppose there are 2 panels, A & B, with efficiency of 17 % and 19% respectively.

If 100 units of sunlight are present on them, panel A will convert 17 units of it to electricity and panel B will convert 19 units of it to electricity.

So, as it is evident from the above, we generally prefer higher efficiency panels.

The solar panels are divided into different generations.

- The first generation solar panels are mostly utilized for installation on residential and commercial scale projects. The material used for such solar panels is silicon and contains:
    a. Monocrystalline solar panels: It is the purest silicon-based panel. Some features of the solar panel include high efficiency, occupation of less

space, reduced deterioration even in harsh environmental conditions and long life.
  b. Polycrystalline solar panels: It is a raw silicon-based panel having some impurities left. Some aspects of the solar panel include reduced efficiency, reduced costs compared to monocrystalline, and its suitability for cold environments since hot temperature have a great adverse effect on them.
- The second generation solar panel contains:
  a. Thin film solar cells: The materials used for these solar panels include copper, cadmium or silicon onto the substrate. The advantages include cheapness and easy production. However, they are still not preferred for installation at residential premises because of reduced lifespan, and reduced efficiency.
  b. Amorphous silicon (A-Si): It contains triple layered technology. The efficiency is also less, but the cost is reduced which is an advantage.
- The third generation solar cell comprises of organic photovoltaics containing:
  a. Dye-sensitized solar cells
  b. Organic material based solar cells
  c. Pervostrikes
  d. Quantum dot solar cells
- The fourth generation solar cell is an emerging PV based on hybrid inorganic crystals with a polymer matrix.

The higher efficiency cells are monocrystalline. This is the latest technology which is rapidly growing. So our preferred choice should be monocrystalline cells.

The second parameter we need to check is the wattage of the panels.

Consider that after our analysis we conclude that the required size for our roof is 5 kW or 5000 W. Let's say we decide to go with a 300 W module.

We will need:

5,000/300 = 17 panels

But what if our roof can only fit 15 panels? If our roof size is small we can't exceed 15 panels!

In such a case, we choose the panels with a higher wattage. Let's see what happens when we choose 350 W modules instead of 300 W.

As a result, we will need:

5,000/350 = 15 panels

But how can one panel produce 300 W and another panel produce 350 W if they are of the same size?

Well, this is where efficiency comes into the picture. Although both the panels are the same size and have the same amount of sunlight present on them, a 350 W panel will have higher efficiency and convert more energy into electricity.

There is one more thing that we need to keep in mind, if a panel is of higher efficiency it will cost more than the panel with lower efficiency.

So, do you really need higher efficiency panels? Yes and no.

Yes, you will need a higher efficiency panel if you are restricted by the roof size and no you won't need them if your roof is large enough.

To elaborate further, if our roof in the above examples was large enough to fit 17 modules we would not need to go for higher efficiency panels (350 W)

To summarize, panel selection depends on roof size more than its efficiency!

Every pilot needs a checklist and so do we!

Checklist:

1. Check how many solar panels can fit on your roof.

2. If the area is less than what is required, increase the wattage and check until you manage to accommodate the required capacity.
3. In case we cannot meet the required capacity try to maximize the capacity with higher wattage panels.
4. Check warranty details before buying from the manufacturers. You would not want to buy panels from a company that could go bust in the coming years! For this specific reason I have mentioned only the most reputable manufacturers.
5. Your solar array system is going to be your asset for the next 20-25 years, so when checking the cost of the system make sure you don't sacrifice quality.

CHAPTER 10:
SOLAR PANEL MAINTENANCE

Figure 10.1 Solar Panel Cleaning

Brushing and flossing our teeth might seem boring at times but it is a necessity. Similarly, going on your roof and cleaning all the solar panels might seem like a difficult and time-consuming task, but it is also a necessity.

Dust, and dirt caused by wind, animals, and birds, can cover the surface area of the solar panels. This results in a decreased output. In order to avoid this, we need to regularly clean the surface area of our solar panels.

Luckily, unlike us, the solar panels require brushing of their surface only **once a month**, given average conditions. If you live in an area where there is a high amount of dust being deposited on your solar panels, then you might need to clean your panels maybe once every two weeks or so. The clearer the panels the higher their efficiency!

CHAPTER 11:
THE INVERTER AND ITS LOSSES

If you recall in the chapter on the basics of solar, we derived a basic block diagram. Something like this:

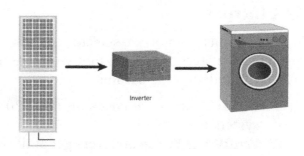

Figure 11.1 Solar PV system connections

The inverter acts as a medium for transferring the power from DC to AC. But in order to do that job it asks for something in return, just as we must pay a fee to a translator, we need to pay, or lose energy in the process of transferring power.

RULE OF THUMB: The inverter has 2% losses!

Damn and I thought only humans were imperfect!

What this means is that if I were to give 100 units of energy to the input, the inverter will give only 98 units as output. However, the datasheets provided by the manufacturers provide efficiency details which shows variation.

- Input= 100
- Output= 98
- Efficiency= 98%
- Losses= 2%

CHAPTER 12: INVERTER SELECTION CRITERIA

12.1 Introduction

In our previous chapter, we selected the solar panels based on certain parameters. In this chapter we are going to select the inverter based on some parameters.

1. **COST:** The inverter cost adds up to 20% to the total system cost.
2. **WARRANTY:** The inverters generally come with a 5-year warranty.
3. **NUMBER OF PHASES:** Single phase vs. three phase.

12.2 What is a Phase?

Let us understand this with the help of an analogy.

Figure 12.1 Analogy of single phase

Imagine you have a tanker filled with water. Since only one person lives in town, there is just one tap. Let that tap be a yellow tap. After a few months two other people come to live in that town. So the government increases the number of taps.

Now this tanker has 3 taps. red, yellow, and blue.

Figure 12.2 Analogy of three phase

Every time anybody needs water they simply connect a pipe to a tap and use the water from the tank. Let's say there are three people P1, P2, and P3 who want water.

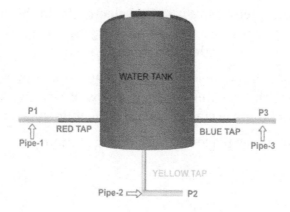

Figure 12.3 Three phase and cable

The three people, P1, P2, and P3 have access to water using pipe-1, pipe-2, and pipe-3 respectively. What if there is a fourth person, P4 who wants access to water?

Where will they get water from?

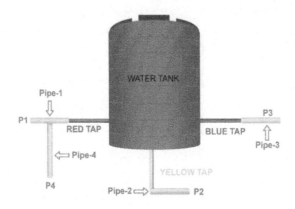

Figure 12.4 Analogy of three phases and joint

As you can see, the fourth person simply connected his pipe to the red tap with the help of pipe 1. Keep this example in mind as we understand phases in electricity.

You must have seen a wire carrying electricity. In this case the wire or the cable is just a single unit carrying power. But what would happen if we needed to deliver more power?

It makes sense that whenever power requirement increases, we prefer to go with a greater number of power carrying wires. Let us understand it more clearly through the following examples:

The following takes place at a generating station:

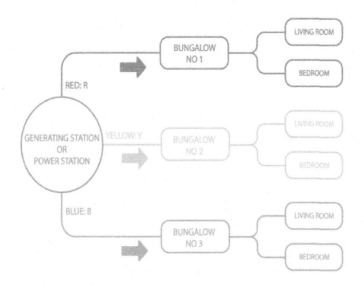

Figure 12.5 Power delivery to the home through three cables

The typical power flow looks like this. The generating station generates power using three different wires or modes or phases. Each phase is connected to a different house. For ease of understanding I have taken just three homes. So, if your area had just three homes and just one three phase generating unit, it would look like the diagram above. **Three taps for three people**. But there are so many houses in your locality, how will you power them using just three wires?

The answer lies with using more wires. **One tap many wires.**

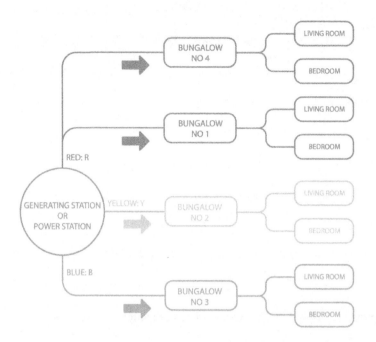

Figure 12.6 Power delivery to the home by taking joints through four cables

As you can see, we have simply added one more red wire. Similarly, we can increase the yellow and blue wires as our requirements increase.

PHASES in layman's terms can be seen as labels assigned to generating energy or **taps to deliver** power. Three outlets of the generator. The three taps of a water tank. Now that you have your own tap, you can connect as many pipes and piping arrangements as you want.

RULE OF THUMB:

A generator always generates three phase power.

12.2.1 What is Single Phase?

Whenever the power requirement in an area is low, a single-phase system is preferred. Two wires enter our home. One phase wire (it can be red, yellow or blue), neutral and earth (more on this a bit later).

Take a look at the following types of systems…

- Line-neutral
- Line-line-neutral (This is also known as two phase)

This can seem a bit confusing when deciding which type of a system to implement in your house. But rest assured, the chapters following this one will make the understanding about phases crystal clear once we reach there. For now, our sole job is to know whether we have a single phase system or a three phase system.

As mentioned in the above figure, the generating unit generates power in three phase and supplies each individual phase to separate areas/ houses.

12.2.2 What is Three Phase?

Whenever the power requirement in any area is more like an industry, manufacturing unit or even residential homes, '3-phase wires are installed (red, yellow, and blue) and sometimes a neutral wire in addition to the above three. (And of course an earth as well). The color code may vary in different countries and with differing local regulations so please refer to the following chart:

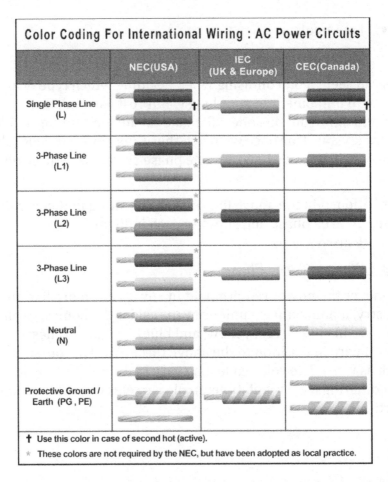

Figure 12.7 Colour Coding for International Wiring: AC Power Circuits

Can a single bungalow/ house get a three-phase connection?

Yes, this can happen. Generally, every individual house gets only one phase connection but in some cases one house might get two or three phases.

Bedroom: red phase

Living Room: yellow phase

Garage: blue phase

Though the majority of the homes get single-phase power, some homes also get three phase power. Some homes have a special air conditioning unit which works on three phase only.

Now that we know the difference between single phase and three phase power, we need to select the inverter in the same manner.

12.2.3 Single Phase or Three Phase

But how would I know if my home is single phase or three phase?

Locate your Main Service Panel (MSP). This is where all the wires running through all your rooms will begin. This is the starting point. The type of system will generally be written on the panel door.

E.g. 240/208V single phase.

415/480V three phase.

Your job is to find these keywords, "240 single phase" or "415 three phase."

If you find 208V/240V it can be considered single phase (also known as two phase).

If you find 415,440V it can be considered three phase.

Now that you know that your home is a three phase system or a single phase system, it's time to finally choose an inverter.

If you have a single phase (or two phase system) system you will need a single phase inverter.

If you have a three phase system you will need a three phase inverter.

The following are a few brands from which you can choose:
- Fronius
- SMA
- Solar Edge
- ABB

12.3 Inverter sizing

The next step in the process is inverter sizing. Just like panels come in different output wattages, inverters too come in different output modes. Let us calculate the inverter size.

Figure 12.7 Inversion operation

If you remember, the output of solar panels is DC and the home appliances require AC. Hence, the inverter converts the DC to AC.

Remember, the inverter can never give more output than is provided at the input.

Let's say we have a rooftop solar array of 10 kW. This would lead us to think that the inverter should be of 10 kW as well. But here's the catch...

RULE OF THUMB:

Even if we install a 10 kW system, it will never give us 10 kW as output. It will always be less.

In our chapter on solar panels and their losses, we saw the various types of losses. These were the direct losses. But there are other hidden losses as well.

12.4 Hidden losses

Generally, 10 am to 4 pm are considered sun hours. But as you can observe, the intensity of the sun's rays is not equal all the time. Therefore, the output will not be at maximum all the time.

Imagine that there is a tall building in your vicinity. This might also block or reduce the intensity of the sunlight.

If you live near a hilly region, these mighty mountains block the sunlight partially or fully at times, reducing the output.

If your house is surrounded by trees, as much as we would love to be in solace within the shadow of them, the truth is, the shadow also blocks the sunlight and reduces the output of our rooftop solar.

So, how do I select an inverter then?
RULE OF THUMB:

The DC power to AC power ratio should be ideally around 1.3 and never exceed manufacturers specs.

Let us understand this in a bit more detail.

Imagine that the output of our rooftop solar array is 10 kW. This is DC power.

If I choose an inverter of 10 kW, the output of this inverter is 10 kW AC power.

So if I were to take a ratio of the DC power to AC power, it would be:

$$\frac{\text{DC Power}}{\text{AC Power}} = \frac{10 \text{ kw}}{10 \text{ kw}} = 1$$

In this case, our DC/AC ratio is **1**.

So we can choose a DC/AC ratio which can go up to **1.3**

If we choose a 7.6 kW inverter for the 10 kW system, our ratio will be:

$$\frac{\text{DC Power}}{\text{AC Power}} = \frac{10 \text{ kw}}{7.6 \text{ kw}} = 1.315$$

This is close to the recommended value. Let us try the next highest size and check the DC/AC ratio

Let us try using an 8.2 kW inverter:

$$\frac{\text{DC Power}}{\text{AC Power}} = \frac{10 \text{ kw}}{8.2 \text{ kw}} = 1.219$$

An 8.2 kW inverter will work perfectly fine but the cost of it will be higher than the 7.6 kW option.

So we can select it as our inverter for a 10 kW system on our roof.

The inverter range starts from as small as 2 kW up to 27.6 kW. When selecting the inverter, we need to check the DC/AC ratio and then select the inverter.

Here are a few Fronius inverter sizes:
- 3 kW
- 3.5 kW
- 3.8 kW
- 4 kW
- 4.6 kW
- 5 kW
- 6 kW
- 8.2 kW

Here are a few Solar Edge inverter sizes:
- 3 kW
- 5 kW
- 6 kW
- 7.6 kW
- 8 kW
- 10 kW

Every manufacturer gives these details in the datasheet. These datasheets are easily available online. We will shortly reach the chapter where I will explain in detail how to read a technical datasheet.

Checklist for selection of inverters:

1. Cost: Before choosing a brand compare the cost, as it is 20% of your total system cost.
2. Warranty: Most of the brands provide a 5-year warranty. Your selected brand will be able to provide the same warranty.
3. Determine whether your home is supplied using single phase or three phase, by following the steps mentioned in this chapter.
4. For a single phase system choose a single phase inverter and for a three phase system select a three phase inverter.
5. Once you know the phase type and the system size evaluate the DC/AC ratio.
$$\frac{\text{DC Size}}{\text{AC Size}} < 1.33$$
6.
7. DC/AC ratio should be around 1.3

Where...

DC size = Size of your solar array in kW.

AC size = Output of the selected inverter. e.g. Fronius 7.6 kW.

Therefore AC = 7.6 kW.

CHAPTER 13: THE BATTERIES

13.1 The Benefits of Using Solar Batteries

Installing a solar battery storage system enables you to get the very most out of your solar energy system. It enables you to store any electricity your panels have generated during the day and can improve the overall performance of your solar energy system.

Introducing solar battery storage into your system brings you closer to full energy independence, and means you're not solely reliant on sunlight to meet your energy requirements at any one time. With a battery storage system you're maximizing your use of your solar energy system and reducing pressure on the grid.

Let's take a look at some of the main benefits of solar battery storage:

Solar battery storage maximizes energy use. An on-grid solar system without a solar battery will draw power from the grid when the sun is not shining, and so there will be a bill to pay for this electricity used.

Installing a solar battery system enables consumers to lower their electricity bills. In addition, any excess electricity can even be sold to the electricity grid via a net-metering mechanism.

Solar battery storage systems help us to reduce our reliance on the grid.

Disadvantages of Solar Battery Storage Systems

One disadvantage of electricity storage devices is their price. The devices are often still too expensive to make their purchase economically worthwhile. Wholesale devices made of lithium-ion storage cells have become significantly cheaper, but not necessarily for consumers. Other storage technologies have played only a marginal role in residential applications for the past decade and are more expensive than lithium-ion storage.

It is unclear how long lithium-ion storage can operate. Since 2011, they have been installed in photovoltaic systems. However, it is not known what condition the oldest devices are now in. The manufacturers are reluctant to make public the empirical values regarding the service life of their products.

Net Metering

Net metering gives owners of solar systems the chance to make money by selling any extra electricity generated to the grid. In addition, any shortfalls in generated electricity can be made up for by buying It from the grid. So if the amount of energy generated is more than is needed, the solar system owners will be paid for that surplus by the electricity company, who will then add that extra electricity into the grid. In the reverse situation, f the amount of electricity consumed is greater than the amount of electricity generated, then the system will draw energy from the utility grid and the owner pays the net amount. As of March 2015, 44 states and Washington, D.C. developed mandatory net metering rules for some utilities.[3] Although the states' rules are well laid out, in reality, very few utilities give compensation at full retail rate levels.[4] Net metering policies are determined by individual states and they have specific policies that vary in several ways. The Energy Policy Act of 2005 requires state electricity regulators to consider that mandate public electric utilities make net metering available to their customers upon request. However, the key word here is 'consider' as it does not actually have to implement these rules. Various legislative bills have been proposed to set a federal standard limit on net metering. These are quite varied, such as H.R. 729, with a net metering cap at 2% of the forecasted aggregate customer peak demand, and H.R. 1945 with no aggregate cap but a limit for residential users of 10 kW. This is a lower limit than many other states, such as New Mexico, which has a limit of 80,000 kW or states that limit as a percentage of load such as Ohio, Arizona, Colorado, and New Jersey.

State	Subscriber limit (% of peak)	Power limit Res/Com(kW)	Monthly rollover	Annual compensation
Alabama	no limit	100	yes, can be indefinitely	varies
Alaska	1.5	25	yes, indefinitely	retail rate
Arizona	no limit	125% of load	yes, avoided-cost at end of billing year	avoided cost
Arkansas	no limit	25/300	yes, until end of billing year	retail rate
California	5	1,000	yes, can be indefinitely	varies
Colorado	no limit	120% of load or 10/25*	yes, indefinitely	varies*
Connecticut	no limit	2,000	yes, avoided-cost at end of billing year	retail rate
Delaware	5	25/500 or 2,000*	yes, but banked kWh expire yearly on 31 Mar.	retail rate
District of Columbia	no limit	1,000	yes, indefinitely	retail rate
Florida	no limit	2,000	yes, avoided-cost at end of billing year	retail rate
Georgia	0.2	10/100	no	determined rate
Hawaii	none [10]	50 or 100*	yes, until end of billing year	none[11]
Idaho	0.1	25 or 25/100*	no	retail rate or avoided-cost*
Illinois	1	40	yes, until end of billing year	retail rate
Indiana	1	1000	yes, indefinitely	retail rate
Iowa	no limit	500	yes, indefinitely	retail rate

State	Subscriber limit (% of peak)	Power limit Res/Com(kW)	Monthly rollover	Annual compensation
Kansas	1	25/200	yes, until end of billing year	retail rate
Kentucky	1	30	yes, indefinitely	retail rate
Louisiana	no limit	25/300	yes, indefinitely	avoided cost
Maine	no limit	100 or 660*	yes, until end of billing year	retail rate
Maryland	1500 MW	2,000	yes, until end of billing year	retail rate
Massachusetts**	6 peak demand 4 private 5 public	60, 1,000 or 2,000	varies	varies
Michigan	0.75	150	yes, indefinitely	partial retail rate
Minnesota	no limit	40	no	retail rate
Mississippi	—	—	—	wholesale rate plus 2.5 cents per kwh standard, plus an additional 2.0 cents for low income customers [12]
Missouri	5	100	yes, until end of billing year	avoided-cost
Montana	no limit	50	yes, until end of billing year	lost [13]
Nebraska	1	25	yes, until end of billing year	avoided-cost
Nevada	3	1,000	yes, indefinitely	retail rate
New Hampshire	1	100/1,000	yes, indefinitely	avoided-cost
New Jersey	no limit	previous years consumption	yes, avoided-cost at end of billing year	retail rate

State	Subscriber limit (% of peak)	Power limit Res/Com(kW)	Monthly rollover	Annual compensation
New Mexico	no limit	80,000	if under US$50	avoided-cost
New York	1 or 0.3 (wind)	10 to 2,000 or peak load	varies	avoided-cost or retail rate
North Carolina	no limit	1000	yes, until summer billing season	retail rate
North Dakota	no limit	100	no	avoided-cost
Ohio	no limit	no explicit limit	yes, until end of billing year	generation rate
Oklahoma	no limit	100 or 25,000/year	no	avoided-cost, but utility not required to purchase
Oregon	0.5 or no limit*	10/25 or 25/2,000*	yes, until end of billing year*	varies
Pennsylvania	no limit	50/3,000 or 5,000	yes, until end of billing year.	"price-to-compare" (generation and transmission cost)
Rhode Island	2	1,650 for most, 2250 or 3500*	optional	slightly less than retail rate
South Carolina	0.2	20/100	yes, until summer billing season	time-of-rate use or less
South Dakota	—	—	—	—
Tennessee	—	—	—	—
Texas***	no limit	20 or 25	no	varies
Utah	varies*	25/2,000 or 10*	varies - credits expire annually with the March billing*	avoided-cost or retail rate*
Vermont	15	250	yes, accumulated	retail rate[14]

State	Subscriber limit (% of peak)	Power limit Res/Com(kW)	Monthly rollover	Annual compensation
			up to 12 months, rolling	
Virginia	1	10/500	yes, avoided-cost option at end of billing year	retail rate
Washington	4.0 percent of the utility's peak demand during 1996	100	yes, until end of billing year	retail rate
West Virginia	0.1	25	yes, up to twelve months	retail rate
Wisconsin	no limit	20	no	retail rate for renewables, avoided-cost for non-renewables
Wyoming	no limit	25	yes, avoided-cost at end of billing year	retail rate

How do Solar Batteries Work?

The basic principle of the battery is based on the conversion of chemical energy into electrical energy by carrying out oxidation and reduction reactions. Such reactions are based on the formation of an electron by carrying out a chemical reaction. There are primary batteries and secondary batteries. In primary batteries, a one-way reactional procedure takes place by carrying out the conversion of chemical energy into electrical energy. The process is irreversible and the electrical energy cannot be converted back into chemical energy. Such batteries are considered to be non-chargeable batteries. An example of such a battery is a common cell that is used in flashlights and small equipment such as television remotes. Another category is the secondary battery which carries out the conversion of chemical energy into electrical energy and vice versa. This allows the batteries to be charged and discharged. Examples of such batteries are lead-acid batteries, nickel-metal hydride batteries, and lithium-ion batteries. Some of the features provided by the EST to the consumer depending upon the scale are provided in the table below.

Table 4 EST and their Applications

Applications	Duration of storage (hours)	The scale of the system (Wattage)
Power delivery to the consumer	1h-10h	Small (Less than 1MW)
Peak shaving, time-shifting, and generation	1h-10h	Large (More than 300MW)
Deferral of transmission and distribution	5h-12h	Large (more than 300MW)
Capacity firming	Hrs.-days	Large (more than 300MW)
Load leveling	Hrs.-days	Large (more than 300MW)
Unit commitment	Hrs.-days	Large (more than 300MW)

The comparative analysis of ESS technology is shown in the figure below. The figure shows five distinct types of technologies i.e., electrical storage technologies, electrochemical storage technologies, thermal storage technologies, mechanical storage technologies, and hydrogen storage technologies. For different types of technologies, the three main characteristics are analyzed, which are reserves and response services, transmission and distribution support, and bulk power management. In the case of battery storage systems (BSS), it can be observed that the BSS systems contribute to reserve and response services, and transmission and distribution support from seconds to hours, but have reduced capability of managing the power at the bulk level. The redox flow batteries have the capability of supporting the transmission and distribution system which may range from kilowatts to megawatts. In order to analyze the behavior of the BSS, the basic characteristics and types of the batteries are discussed in the below sections

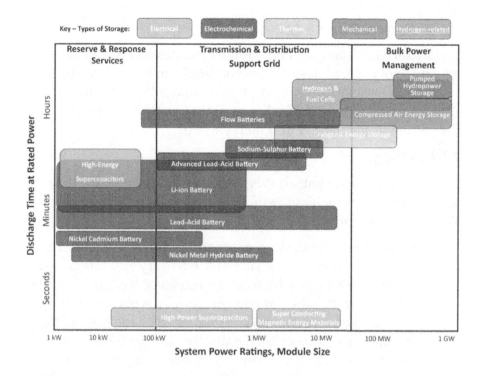

Figure 13.1 Battery systems

13.2 Battery Characteristics

13.2.1 Battery Requirements

When it comes to installation purposes, the requirements of the battery systems are as follows:
- a. The battery must have a long life
- b. The installation and maintenance costs should be low
- c. The charge storage efficiency should be high
- d. The self-discharge capacity of the battery should be low

13.3 Battery Basics

Before understanding all these factors, it is necessary to understand the different terms associated with batteries.

13.3.1 Battery Cells and Packing

The battery pack contains individual modules as well as cells which are organized in series and parallel combinations. The cells of the battery range from 1V to 6V and different cells are connected in series and parallel combinations which are then assembled in the battery and then packed.

13.3.2 Efficiency

The efficiency of the batteries is the first important factor that affects the cost as well as the lifetime of the photovoltaic system. The three efficiencies are:

1. **Charge efficiency:** It is an efficiency that is measured at the constant discharge rate of the battery and refers to the amount of charge which can be retrieved from the battery with respect to the amount the battery is placed to charge. in other words, it can be considered to be the ratio of the number of charges which enter the battery while the charging process is carried out, when compared with the number of charges being extracted from the battery while the discharging process is carried out. It is to be noted that self-discharge. being a deteriorating factor, will reduce the coulombic or charge efficiency of the battery. Generally, the coulombic efficiency can be as high as 95%.

$$\eta_{col} = \frac{Q_{discharging}}{Q_{charging}} \times 100$$

2. **Voltage efficiency:** This efficiency provides a difference in the voltage i.e., charging voltage and discharging voltage. This factor is also measured at a constant discharge rate, but it reflects the way the charge is retrieved at the voltage contrary to that of the charging of the battery.

$$\eta_{vol} = \frac{V_{discharging}}{V_{charging}} \times 100$$

3. **Energy efficiency:** This is the product of charge efficiency and voltage efficiency.

$$\eta = \frac{E_{out}}{E_{in}} \times 100$$

This can be illustrated by an example of getting the

$$\eta = \frac{9kWh}{10kWh} \times 100 = 90\%$$

13.3.3 Power Rating

The power rating of a battery is measured in amperes (A) and can be defined as the maximum amount of charge and discharge that the battery has.

13.3.4 Battery Capacity

The battery capacity defines the highest level of energy that is extractable from the battery while the voltage of the battery does not fall below the described value by the manufacturer. It is measured in ampere-hours (Ah) or kilowatt-hours (kWh). In other words, the capacity of the battery defines the charge which can be delivered by the battery's rated voltage and is directly proportional to the amount of electrode material utilized in the battery. This value varies depending upon the type of the battery. A notable factor is that the capacity of the battery falls by approximately 1% for every 20°C.

13.3.5 Depth of Discharge (DoD)

The DoD defines the percentage of the electric capacity of the battery which can be withdrawn from the battery. The battery's DoD level range is between 25% to 80% of the rated capacity. This can be illustrated by an example that the battery has the capacity of 10Ah when drained out by 2Ah, then the DoD of the battery remains 20%. The DoD and SoC are interrelated to each other and can be obtained by taking complement among them.

$$DOD = \frac{E_{discharged.}}{C_{battery} \, V} \times 100$$

13.3.6 State of Charge (SoC)

The SoC defines the percentage of the electric capacity of the battery which is available for discharging purposes. This can be illustrated by an example that the battery has the capacity of 10Ah when drained out by 2Ah, then the SoC of the battery remains (10-2=8) 80%. The battery's DoD level range is between 25% to 80% of the rated capacity.

$$DOD = \frac{E_{discharged}}{C_{battery} \, V} \times 100$$

13.3.7 C-rate

The C-rate of the battery defines the rate of the charge at which the battery is being discharged with respect to its capacity. The C-rate can be further illustrated by an example. the C-rate of 1C for the battery capacity of 10Ah corresponds to the discharging of the current having value of 10A over one hour. Similarly, if the C-rate is 0.5C, it means that the battery will have a discharge current of 5A over 2 hours.

$$C - rate = \frac{I}{C_{batt}/hour}$$

Or,

$$I = C - rate \times C_{batt}/hour$$

Using above mentioned example

$$I = 1 \times 10Ah/1\,hr = 10A$$

$$I = 0.5 \times 10Ah/1\,hr = 5A$$

Similarly, the same formula can be utilized to find hours and the C value of the C-rate.

13.3.8 Lifecycle

The life cycle of the battery is the most important factor and can be defined as the number of cycles of charging and discharging after which the capacity of the battery falls by 80%. It is mostly affected by the way the battery is being charged and discharged, and the temperature of the environment in which the battery is placed. Therefore, at lower temperatures, the battery capacity remains lower. The reason is that at higher temperatures the chemicals remain highly active. But, the high temperature is also detrimental to the health of the battery.

13.3.9 The Lifespan of Storage Devices

The end of the service life of electricity storage devices is defined by the electrical and storage industry in terms of storage capacity. As with any other storage device made of lithium-ion cells (e.g. in a cell phone or notebook), this capacity decreases over time as a result of aging. A distinction is made here between the cycle stability of the battery and its calendar life. The cycle stability indicates how many charge and discharge cycles the battery can undergo during operation. The calendar life refers to the aging of the materials used in the battery, which wear out and decay over time. If the storage capacity according to the nameplate (= nominal capacity) falls below 80%, this is usually considered the end of life for lithium-ion batteries in stationary applications. This does not necessarily mean that the battery can no longer be used. But the "first life" of the battery is then over; a so-called "second life" may still follow. It is up to the operator to decide whether and how the use of a battery with a greatly reduced storage capacity is still sensible.

13.3.10 Specific Power

The specific power defines the availability of maximum power per unit mass of the battery. It is associated with characteristics of the chemistry and packaging of the battery and is measured in watts per kilogram (W/kg). It helps in the determination of battery weight which will be required for the achievement of performance.

13.3.11 Energy Density

The energy density can be thought of as the nominal energy of the battery per unit volume. It is also referred to as the volumetric energy density of the battery. It is also associated with characteristics of the chemistry and packaging of the battery and is measured in watt-hours per liter (Wh/L). It helps in the determination of battery weight which will be required for the achievement of performance.

13.3.12 Power Density

The power density is the nominal power of the battery per unit volume. It is also referred to as the volumetric power density of the battery. It is also associated with characteristics of the chemistry and packaging of the battery and is measured in watt-hours per liter (W/L). It helps in the determination of battery weight which will be required for the achievement of performance.

13.3.13- Manufacturers and Performance Warranties

The experience and reputation of each battery manufacturer can also make a difference. A well-established and popular brand is likely to provide you with the most reliable products and the best customer service.

13.4 Types of Batteries

The battery storage systems comprise different types of batteries. The most commonly utilized batteries are lead-acid batteries, lithium-ion batteries, nickel-metal hydride batteries, and vanadium redox flow batteries. In the below sections the batteries, along with their characteristics will be discussed.

13.4.1 Lead Acid Battery

The lead-acid battery is one of the oldest batteries and has been used as storage technology in both residential and industrial sectors since the 1900s. The construction of the battery such that the two electrodes i.e., the positive electrode and negative electrode of the battery utilize metallic lead and lead oxide, that is immersed in the solution of dilute sulfuric acid which acts as an electrolyte.

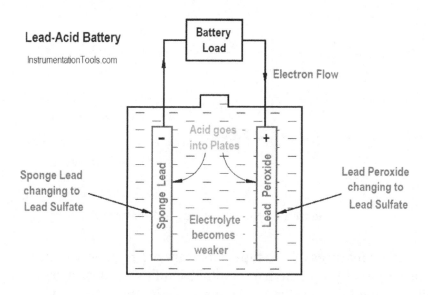

Figure 13.2 Lead-Acid Battery

The lead-acid batteries are further divided into two main categories which are sealed valve-regulated lead (VRLA) acid and flooded lead-acid batteries.

The advantage of lead-acid batteries is their reduced capital cost, high energy efficiency, and a lower rate of self-discharge which is usually less than 20%. Such batteries are usually used on a residential scale or in commercial applications which take into account the economics. However, the energy density and lifetime of the battery are less than that of other batteries.

13.4.2 Nickel Metal Hydride (NiMH) Battery

The nickel metal hydride battery comprises two different materials for the electrodes. the positive electrode is made up of nickel oxyhydroxide, and the negative electrode comprises cadmium. Potassium hydroxide is used as an alkaline electrolyte.

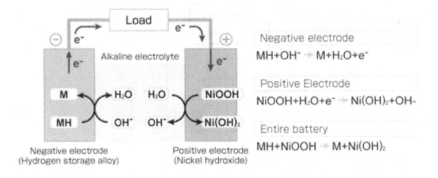

Figure 13.3 Nickel metal hydride battery

The main features of this type of battery are high power and energy density, and environmental friendliness. However, limitation of the service life, reduced coulombic efficiency (approx. 65%), and a high rate of self-discharge are some limitations of this type of battery. The applications of the battery include their use in electric vehicles such as Chevrolet Malibu hybrids, Toyota Prius, Ford Escape, and Honda Insight.

13.4.3 Lithium-ion (Li-ion) Battery

Lithium-ion batteries are characterized by:

- a high storage density,

- high discharge currents,

- high efficiency and

- high cycle stability.

The first two points are primarily important for use in electric cars. The batteries can be comparatively small and light and enable fast charging. High efficiency is advantageous for any application. High cycle stability is an important feature for stationary storage devices as an electricity storage unit goes through around 250 cycles per year. This means that it is charged with solar power about 250 times, which is then discharged again. If the storage device runs for 15 years, this results in around 4,000 charging cycles; if it remains in operation for 20 years, it will go through 5,000 charging cycles in its lifetime. This is not a problem for lithium-ion batteries - they can manage 10,000 charging cycles. Nevertheless, the storage capacity slowly decreases during the course of operation. The storage capacity can also decrease due to calendar aging. You should therefore not rely solely on the specified number of cycles when buying a battery. The battery can reach the end of its calendar life before it has gone through 4,000 or 5,000 charging cycles.

However, there are two points that are repeatedly criticized in connection with lithium-ion storage systems: The mining of the metals used in the storage cells - above all lithium and cobalt - is in some cases associated with environmental destruction. And the fact that the chemicals used are flammable and explosive. There have been cases in Germany where lithium-ion storage batteries have set on fire. It's not the norm, but it happens. It is important that a battery is operated within its defined range. To do this, there is a battery management system in the storage unit that monitors and controls the temperature of the storage cells, the charging and discharging currents, and the electrical voltage of the individual storage cells. The battery management system also ensures that the storage unit is not completely discharged, and that the depth of discharge (DoD) does not fall below 90%, so that the battery's service life is not damaged. It is important for the user to know: A storage unit should be installed in a cool place where there are no high temperature differences over the year. Ideal temperatures are 15 to 20 degrees. Above 25 degrees the cells age more quickly. In addition, the battery should not be kept fully charged for a long time.

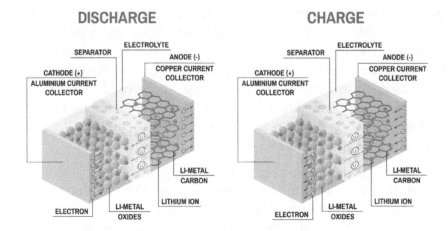

Figure 13.4 Lithium-ion battery

13.4.5 Vanadium Redox Flow (VRB) Battery

The redox flow batteries (RFB) were developed by the U.S National

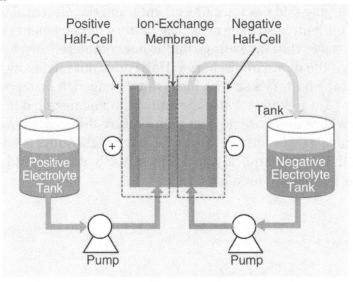

Figure 13.5 Vanadium redox flow battery

Aeronautics and Space Administration in the 1970s. Different types of batteries have been developed - iron chromium-based

RFB batteries, zinc/bromine hybrid flow batteries, polysulfide/bromine batteries, and vanadium redox flow batteries.

Among different categories, the vanadium redox flow battery has proved to be the most prominent RFB. The two distinct features of

this battery is (i) employment of the same material in both positive and negative electrolytes, and no excessive generation of hydrogen. The figure shown above illustrates the main principle of RFB. Unlike other batteries, this type of battery contains a separate power module and energy module. The power module carries out conversion of electrical and chemical energy. The energy module comprises of an electrolyte tank. Pumping is required for carrying out energy conversion. Therefore, the electrolyte in liquid form is pumped from the tanks to the cell resulting in the occurrence of a chemical reaction at the electrode.

13.4.6 Comparative Analysis of Batteries

The comparative analysis of the batteries is provided in the table to analyze all the main factors which help in choosing the batteries while considering technical and economic aspects.

Table 5: Comparative Analysis of Batteries

Factors	Units	Battery Type							
		Lead acid		Li-ion		NiMH		VRB	
		Low	Up	Low	Up	Low	Up	Low	Up
Energy density	Wh/kg	25	50	75	200	60	120	10	30
Power density	W/kg	75	300	500	2000	250	1000	80	150
Life cycle	100% DOD	200	1000	1000	10000	180	2000	~12000	
Capital cost	$/kWh	100	300	300	2500	900	3500	150	1000
Efficiency	%	75	85	85	97	~65		75	90
Self-discharge		Low		Medium		High		Negligible	

The chart presented above illustrates the fact that the lithium-ion battery has considerably higher power density and energy density when compared with other types of battery. Moreover, the life cycle of the battery is quite high when compared with lead-acid batteries and NiMH batteries. Similarly, the cost of the battery is considerably high when compared with lead-acid batteries but relatively low when compared with NiMH batteries. Contrary to that, the self-discharge capability of VRB batteries is negligible and the life cycle is highest when compared with other types of battery. These two factors make VRB batteries the most suitable, but the two main parameters of this battery, which are energy density and power density, are quite low compared to every type of battery which resists its commercialization at residential scale or even commercial scale.

Design of Solar Battery Storage Systems

An electricity storage unit should match your electricity needs. A rule of thumb says: The usable storage capacity of the device should be a maximum of 1.5 times as large as your electricity demand per year. So, if your electricity demand is 5,000 kilowatt hours per year, then the storage unit should have a maximum of 7.5 kilowatt hours of storage capacity. However, a smaller device can also be chosen.

Usable Storage Capacity

There are usually two storage capacities listed on battery storage data sheets; the full storage capacity and the usable storage capacity. Since the devices are not allowed to discharge completely and the integrated charge management prevents this, the full capacity is not available during operation - only the usable capacity. This is therefore somewhat smaller than the full storage capacity, namely by the depth of discharge (usable storage capacity = storage capacity x depth of discharge). So the usable capacity is the size that is important when choosing a storage device.

The storage is there to buffer some of the solar power you generate, for you to use when generation drops. If you can only partially retrieve the stored electricity, the storage will have less free capacity for new solar electricity the next day. Therefore, you can also determine what your electricity consumption is from evening to the next morning. Either read the electricity meter in the evening and morning or call up the consumption values on the meter if it is a digital one. You can base the storage capacity on the electricity consumption in the evening and night hours.

Some manufacturers offer modular storage units. These devices have a control module that can be connected to several storage modules. If you initially buy only two storage modules and there is still room for three more, you can purchase the others at a later date and have them installed in the storage cabinet without much effort, for example if you want to store more electricity or your electricity consumption increases. Another advantage of modular devices is that a single module can be replaced if it is defective.

Storage Clouds

A cloud package can also be used for storage. This means that any electricity that you neither use directly nor store, flows into the power grid and this electricity is credited to a cloud account. This happens mainly in summer when the PV system has its highest yields. On days when the plant generates less power, generally in the fall and winter, you can draw power from the cloud. This is not physical storage, but virtual. The cloud provider takes the electricity fed into the grid and markets it. If you access your cloud account, the cloud provider buys electricity on the market and delivers it to you. The cloud provider thus becomes your electricity supplier for the quantities of electricity you draw from the public grid.

CHAPTER 14:
How to Install Batteries?

The practical implementation of the batteries in terms of installation is based on three combinations which are:

1. Series combination
2. Parallel combination
3. Series-Parallel combination

The way in which the battery connection is being carried out depends upon the requirement of voltage and current. These configurations are defined below:

14.1.1 Series Combination

The series combination is based on connecting the positive terminal of the battery with the negative terminal of the battery, and then the negative terminal of the battery to the positive terminal to form the series configuration. It is to be noted that the current in the series connection remains the same, and the voltages are added. The equation for this is shown below.

$$V_{total} = V_1 + V_2 + \cdots + V_n$$

The technical illustration of the connection is shown in the figure below.

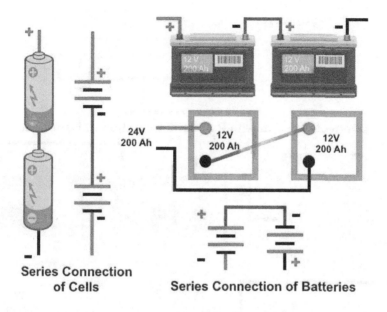

Figure 14.1.1 Series connection of batteries

This can be illustrated by the fact that the battery is of 200Ah, and the voltage is 12V. So:

$$V_{total} = 12V + 12V = 24V \ (at \ 200Ah)$$

14.1.2 Parallel Combination

The parallel combination is based on connecting the positive terminal of the battery with another positive terminal of the battery, and then the negative terminal of the battery to another negative terminal to form the parallel configuration. It is to be noted that the voltage in the parallel connection remains the same, and the currents are added. The equation for this is shown below.

$$I_{total} = I_1 + I_2 + \cdots + I_n$$

The technical illustration of the connection is shown in the figure below.

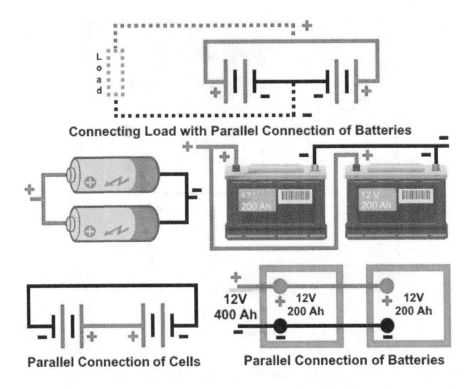

Figure 14.1.2. Parallel connection of batteries

The equation is provided below.

$$I_{total} = 200 + 200A = 400A, (at\ 12V)$$

14.1.3 Series-Parallel Combination

The series-parallel combination is based on connecting the two batteries in series and then connecting them in a parallel combination to make a series-parallel combination. It is a complex circuit, and the circuit is neither totally in series or parallel combination.

Assuming batteries 1 and 2, 3 and 4, 5 and 6 are connected in series, the combinations will have 12V each and have a current of 200Ah which is illustrated below:

B1 and B2

$$V_{B1+B2} = 12V + 12V = 24V\ (at\ 200Ah)$$

B3 and B4

$$V_{B3+B4} = 12V + 12V = 24V \ (at\ 200Ah)$$

B5 and B6

$$V_{B5+B6} = 12V + 12V = 24V \ (at\ 200Ah)$$

Now, connecting the batteries in parallel. There will be two sets as shown in the figure below. The first set will comprise of B1, B3 and B5. The second will comprise of B2. B4, and B6.

Now, the currents become

$$I_{total} = 200 + 200 + 200 = 600Ah$$

Also, the voltage will be 24V

$$V_{set1+set2} = 12V + 12V = 24V \ (at\ 600Ah)$$

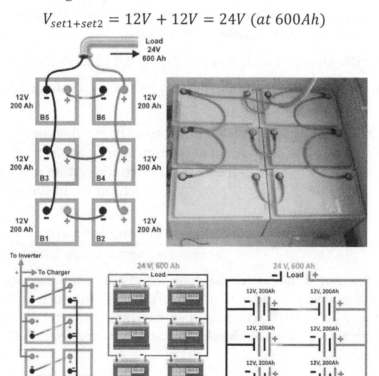

Series-Parallel Connection of Batteries

Figure 14.1.3 Series-Parallel connection of batteries

The technical illustration of the connection is shown in the figure below.

14.2 Battery Sizing for PV System

The battery sizing for the PV system is not very complex. The three main considerations for the load battery sizing are:

1. The nominal voltage of the battery
2. Days of autonomy
3. Total watt-hour requirement of the appliances per day

The steps for the calculation of the battery sizing are:

1. Calculation of watt-hours per day required by the devices, appliances, or equipment

2. Division of watt-hours per day of the battery by 0.85 considering the battery loss of 15%

3. Division of watt-hours per day by 0.6 to account for DoD

4. Division of watt-hours per day by nominal voltage of the battery

5. Multiplication of the whole answer by the days of autonomy. The days of autonomy represent the total number of days of the year in which the system will not be provided with solar power. So, the formula for the sizing of the battery is:

$$Capacity_{battery}(Ah) = \frac{Watt - hrs\ per\ day \times days\ of\ autonomy}{Nominal\ battery\ voltage \times 0.85 \times 0.6}$$

It is to be noted that factors such as battery loss, and DoD are variable and so may depend upon different factors such as the efficiency provided by the manufacturer (at defined temperature conditions) and the way the charges are extracted from the battery.

14.3 Conclusion

In this chapter, the focus was to understand the installation method of batteries. The chapter focused on the basic terminologies of batteries that help in understanding the battery storage systems, and then the characteristics and other related aspects were analyzed. A study was also carried out on the basic chemistry of batteries of different types. The comparative analysis provided the technical and economic evaluation. Moreover, the chapter covered the connection methods along with illustrations and characteristics covering series, parallel and series-parallel combination. Lastly, the battery sizing method was discussed for solar PV systems. The chapter emphasized the battery connections while focusing on the empowerment of renewable energy sources by enhancement of reliability, access to energy in remote areas, and contribution to storage technologies.

CHAPTER 15: READING TECHNICAL DATASHEETS

Reading novels is fun, but reading technical datasheets can seem like a daunting task. The billionaire investor Warren Buffet reads around 500 pages a day before selecting his stocks, luckily for us we just need to read 3-4 pages and that's just once, before selecting our panels and inverters.

15.1 Reading solar panel datasheets:

STEP 1: Open your web browser and search for the panel datasheet.

In this case I am going to select REC 350 Module. Go ahead and type *"REC 350W datasheet pdf"*.

STEP 2: Open any website of your choice and search for "Manufacturer datasheet". Some websites directly provide us with the pdf.

STEP 3: Go to page number 2 and jump to electrical data @ STC.

ELECTRICAL DATA @ STC	Product Code*: RECxxxTP2S 72					
Nominal Power - P_{MPP} (Wp)	330	335	340	345	350	355
Watt Class Sorting - (W)	-0/+5	-0/+5	-0/+5	-0/+5	-0/+5	-0/+5
Nominal Power Voltage - V_{MPP} (V)	38.1	38.3	38.5	38.7	38.9	39.1
Nominal Power Current - I_{MPP} (A)	8.67	8.75	8.84	8.92	9.00	9.09
Open Circuit Voltage - V_{OC} (V)	46.0	46.2	46.3	46.5	46.7	46.8
Short Circuit Current - I_{SC} (A)	9.44	9.52	9.58	9.64	9.72	9.78
Panel Efficiency (%)	16.5	16.7	16.9	17.2	17.4	17.7

Values at standard test conditions STC (airmass AM 1.5, irradiance 1000 W/m², cell temperature 77°F (25°C)).
At low irradiance of 200 W/m² (AM 1.5 and cell temperature 77°F (25°C)) at least 95% of the STC module efficiency will be achieved.
*xxx indicates the nominal power class (P_{MPP}) at STC, and can be followed by the suffix XV for modules with a 1500 V maximum system rating.

Figure 15.1 Datasheet of REC 350W datasheet

Let us understand each and every term and their meanings.

Let us study the important electrical parameters.

15.1.1 Nominal Power (Wp):

Nominal Power - P_{MPP} (Wp)	330	335	340	345	350	355

Figure 15.2 Nominal power of solar panel

Until now, in this chapter we have been talking about panel sizes in terms of their wattages. The nominal power is nothing but the variety of wattages in which the solar panels are available. If you take a look at the datasheet above, REC shows 330W, 335W, 340W, 345W, 350W, and 355W. We have already seen how panel wattage impacts the selection of the panels.

15.1.2 Watt Class Sorting:

This is the tolerance provided by the manufacturer. If you observe, there are two digits and two signs.

Watt Class Sorting - (W)	-0/+5	-0/+5	-0/+5	-0/+5	-0/+5	-0/+5

Figure 15.3 Tolerance level of solar panel

Positive Sign: (+5) This means that a 330W rated panel might give slightly more output which can be up to 5W. Hence, the 330W panel can give out anywhere in the range from 330W to 335W.

Negative Sign: (-0) This means that the 330W rated panel will have zero negative tolerance. Therefore, the output of this panel will not go below 330W. All the panels nowadays provide zero negative tolerance. Up until a few years ago some manufacturers sold panels with a negative tolerance.

Similarly, the output of a 340W rated panel can lie in the range of 340W to 345W and not below 340W.

15.1.3 Nominal power voltage (Vmp)

Nominal Power Voltage - V_{MPP} (V)	38.1	38.3	38.5	38.7	38.9	39.1

Figure 15.4 Nominal voltage of solar panel at maximum point

This is the output voltage of the panel under maximum efficiency. This efficiency is arrived at when we connect the solar panels to the inverter in a proper manner. More on this in the chapter on connections. Until now we were just dealing with output power in terms of wattages. But as we have seen previously, electrical power is made up of two components: current and voltage.

So I will get 38.1V at the output of a 330W module. Similarly, I will get 38.3V at 335W and so on.

15.1.4 Nominal power current (Imp)

Nominal Power Current - I_{MPP} (A)	8.67	8.75	8.84	8.92	9.00	9.09

Figure 15.5 Nominal current of solar panel at maximum point

Similar to voltage, this is the output current of the panel under maximum efficiency. The output current for a 330W panel under maximum efficiency is 8.67 amps while that of a 335W panel is 8.75 amps under maximum efficiency.

15.2 What is this connection efficiency that we are talking about?

A solar PV system performs at its peak when the connections to the inverter are done in a calculated manner. Think of it like this, we know that the inverter transforms power, so it will also have some conditions to be fulfilled at its input to give a greater output, a greater efficiency. Basically, **Vmp** and **Imp** are nothing more than the satisfying of input conditions of the panel with respect to the inverter.

15.2.1 Open circuit voltage (Voc)

Open Circuit Voltage - V_{OC} (V)	46.0	46.2	46.3	46.5	46.7	46.8

Figure 15.6 Nominal power of solar panel

This is the output voltage of the panel under no load condition. If you'd just got back the shop, taken the panel out in the sunlight and connected it to a voltmeter in parallel, you would get this value. Since there is no device connected to it, the panel has no problem in supplying a higher voltage. Think of it as this way:

Imagine a man carrying a light load vs. a heavy load.

As you can see, there is a reduction in the height of the person due to the weight.

Similarly, when there is no weight on the panel i.e. there are no devices/appliances connected to the output of the panel, the panel can supply a higher voltage. whereas when devices are connected to its output, the voltage decreases. The Vmp is less than the Voc as Vmp is the condition wherein the load/device is connected to the output side.

Figure 15.7 Illustration: Load impact on system

15.2.2 Short circuit current (Isc)

| Short Circuit Current - I_{sc} (A) | 9.44 | 9.52 | 9.58 | 9.64 | 9.72 | 9.78 |

Figure 15.8 Short circuit current

Every device/ piece of equipment has a limit. In electrical circuits, this limit is known as the short circuit current. If we keep connecting more and more devices at the output it will result in the drawing of more and more current from the panel. Eventually there will be a point when we reach a limit after which the panel will catch fire and/or it won't be able to supply more current. This limit is the short circuit current.

This can be seen as the overload condition.

What will be the voltage in the case of a short circuit?

Imagine that same man in the previous example was told to carry a huge amount of weight. Something like 500 pounds maybe. We can't be sure of how many bones he might break but he will surely fall down and his height will be zero.

Similarly, the voltage in the case of a short circuit is zero!

15.2.3 Efficiency

| Panel Efficiency (%) | 16.5 | 16.7 | 16.9 | 17.2 | 17.4 | 17.7 |

Figure 15.9 Solar panel efficiency

This is the efficiency in terms of the amount of light energy we are able to convert into electrical equivalent. For a 330W panel we can convert only 16.5% of total light energy into electrical energy. Similarly, for a 355W panel we can convert 17.7% of total light energy into electrical equivalent.

What is STC?

ELECTRICAL DATA @ STC	Product Code*: RECxxxTP2S 72

STC: Standard Test Conditions

What are these standard test conditions?

Values at standard test conditions STC (airmass AM 1.5, irradiance 1000 W/m², cell temperature 77°F (25°C).
At low irradiance of 200 W/m² (AM 1.5 and cell temperature 77°F (25°C)) at least 95% of the STC module efficiency will be achieved.
* xxx indicates the nominal power class (P_{MPP}) at STC, and can be followed by the suffix XV for modules with a 1500 V maximum system rating.

Figure 15.10 Standard Test Conditions

These conditions contain various parameters such as air mass, ambient temperature, cell temperature, irradiance etc. We need to focus just on ambient temperature and irradiance.

Ambient temperature: This is the surrounding temperature where the panels are placed.

The ambient temperature is generally 20-25 °C below cell temperature.

- Ambient temperature= cell temperature- (20-25 °C range)

So at STC, cell temperature= 25 degrees Celsius.

Therefore, ambient temperature= 25-(20-25 range) = (0-5) degrees Celsius.

Irradiance: This is a measure of the amount of energy or the intensity of sunlight.

1000 W/m² implies that on one square meter area 1000 watts of power is incident.

All the above-mentioned electrical characteristics from wattage to short circuit current are valid under **0-5 °C of ambient**

temperature and a minimum irradiance of **1000 W/m²** **irradiance**.

But what if my temperature is different to that and what is irradiance?

ELECTRICAL DATA @ NOCT		Product Code*: RECxxxTP2S 72				
Nominal Power - P_{MPP} (Wp)	244	252	257	260	264	268
Nominal Power Voltage - V_{MPP} (V)	34.9	35.5	35.7	35.8	36.0	36.2
Nominal Power Current - I_{MPP} (A)	6.99	7.10	7.19	7.25	7.32	7.39
Open Circuit Voltage - V_{OC} (V)	42.3	42.8	42.9	43.1	43.2	43.3
Short Circuit Current - I_{SC} (A)	7.44	7.74	7.79	7.84	7.90	7.95

Nominal cell operating temperature NOCT (800 W/m², AM 1.5, windspeed 1 m/s, ambient temperature 68°F (20°C)).
*xxx indicates the nominal power class (P_{MPP}) at STC, and can be followed by the suffix XV for modules with a 1500 V maximum system rating.

Figure 15.11 Electrical datasheet

The electrical data for NOCT has the following conditions:

1. Irradiance= 800 W/m².
2. Ambient temperature= 20 °C.

If you observe carefully, the power output has dropped with the drop in irradiance.

Similarly, the power output has decreased with the rise in temperature. Other parameters also vary with temperature and irradiance.

If you live in an area with 800 W/m² irradiance and a temperature of 20°C, a 330 W panel will give an output of 224 W. Add to that the losses that we have discussed in the previous chapter!

What's the point in installing solar if I am getting less output?

Firstly, while it's true that the solar output falls with increasing temperature and decreasing irradiance, this also happens with all the appliances that we buy. For example, compare the number of months your phone worked flawlessly before it started to slow down.

But it was not supposed to slow down right?

Luckily a good design engineer knows these things and makes the adjustments appropriately. For larger power plants in industries, these factors are extremely important. But when we talk about an individual house, they are taken care of due to the following reasons...

1. Our appliances do not consume power continuously!

If you recall, we started this design process with the example of the washing machine.

Do we connect the washing machine for 24 hours?

Okay, how about the TV, air conditioner, laptop, heater, or water pump?

As you can see, the nature of these devices is irregular. They do not consume power continuously. Hence, the design made using the conditions at STC does match our requirements! At most they would be facing a marginal loss!

But, if we were to design the rooftop array using conditions at NOCT, that would result in a higher number of panels, more space on your roof and a higher investment!

For example:

If we used 15 panels of 350W we would be getting an output of:

ELECTRICAL DATA @ NOCT	Product Code*: RECxxxTP2S 72					
Nominal Power - P_{MPP} (Wp)	244	252	257	260	(264)	268
Nominal Power Voltage - V_{MPP} (V)	34.9	35.5	35.7	35.8	36.0	36.2
Nominal Power Current - I_{MPP} (A)	6.99	7.10	7.19	7.25	7.32	7.39
Open Circuit Voltage - V_{OC} (V)	42.3	42.8	42.9	43.1	43.2	43.3
Short Circuit Current - I_{SC} (A)	7.44	7.74	7.79	7.84	7.90	7.95

Nominal cell operating temperature NOCT (800 W/m², AM 1.5, windspeed 1 m/s, ambient temperature 68°F(20°C).
* xxx indicates the nominal power class (P_{MPP}) at STC, and can be followed by the suffix XV for modules with a 1500 V maximum system rating

Figure 15.12 Variation of results

Ideal= 350 x 15= 5,250W.

Actual= 264 x 15= 3,960W.

So if we were to go for 5,250W as per our initial design, we would need:

5,250/264

= 20 panels.

So we would need an additional 5 panels!

Do I need to redesign everything?

No, we don't need to!

As we have discussed, due to the inconsistency of the consumption in a residential household the design using the STC conditions are sufficient for a residential project.

Going with NOCT conditions will require more roof space, more panels (5 in this case), and more investment!

The problem with this will be a lower rate of return on our investment.

The increase in the output will be less, compared to the increase in the costs!

The increase in savings in doing so, will be less as compared to the increase in investment!

Therefore it doesn't make very good investment sense.

So, finally, we are on track!

Besides, the calculation of 1 watt of solar generating 0.0045 units is considering all the losses. Rest assured, solar won't let you down.

Let's move to the other section of the datasheet where we take a look at the mechanical aspects of a panel.

The information you need now is at the top of page 2.

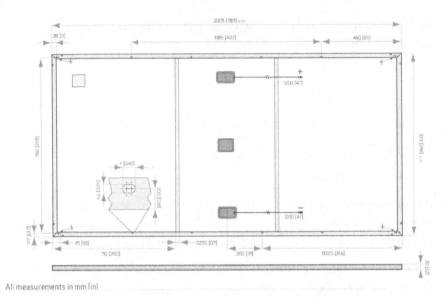

These are the exact dimensions of a solar panel. If you remember in the chapter on sizing, we roughly took the dimensions to be 2m x 1m. Let us verify this.

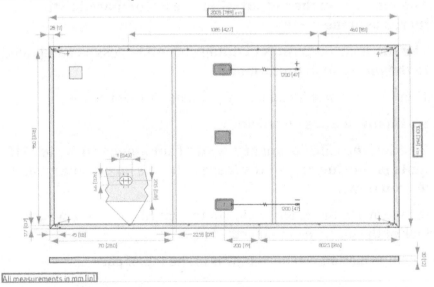

I have highlighted three important things in the above diagram:
1. Units of measurement.
2. Length of the panel.

3. Width of the panel.

Unit of measurement:

All the units of measurement are either in mm or inches. The measurements mentioned in mm are directly mentioned. The units mentioned in inches are in brackets [in]. Let us understand this more clearly.

Length of the panel:

The length of the panel is given at the top of the image.

- Length in mm = 2005 mm ~ 2 m (6.56 ft)
- Length in inches= 789 in

Width of the panel:

The width of the panel is given at the right side of the image.

- Width in mm = 1001 mm ~ 1 m (3.28 ft)
- Width in inches= 394 in

So our assumption was correct, we can generalize the length and width of the panel to 2m x 1m.

CHAPTER 16: READING INVERTER DATASHEETS

We are beginning this chapter with one question in our mind.

How do I connect my 15 (350W x 15 panels= 5kW) solar panels to the inverter?

> **STEP 1:** Identify the type of the system in your house, three phase or single phase. Navigate to your home meter.

240v, 208v single phase, two phase.

440v, 3 pole three phase.

STEP 2: Download the datasheet for the system.

If three Phase, search for Fronius Symo.

If one phase/two phase, search for Fronius Primo.

At the end of the search also enter your country name.

E.g. Fronius primo USA

Before we dive into the inverter datasheet let us take a moment to understand a few key terminologies which will allow us to understand the inverter datasheet more clearly.

- **What is MPPT?**
- **What is MPPT Range?**

What is MPPT?

MPPT refers to maximum power point tracking. Maximum power point refers to the particular voltage (Vmp) and current (Imp) of PV modules which give the maximum output. The process of obtaining this voltage and current from the PV modules is known as maximum power point tracking.

What is MPPT range?

The solar panels are connected to the inverter. So the voltage of the solar panels should lie within a given range to ensure maximum efficiency. This range is called the MPPT range.

For now, we can infer that for optimum conditions we need to follow the conditions of MPPT. The output of the solar panels should lie within the MPPT range of the inverter. This will become clearer as we understand the inverter.

Select the inverter based on our rule of thumb of DC/AC ratio= 1.3

If you recall, we had selected the Fronius 3.8 kW inverter. So let's continue with the same example.

Jump to Page 2 of the inverter datasheet which will look like this:

TECHNICAL DATA FRONIUS PRIMO

INPUT DATA		PRIMO 3.8-1	PRIMO 5.0-1	PRIMO 6.0-1	PRIMO 7.6-1	PRIMO 8.2-1
Recommended PV power (kWp)		3.0 - 6.0 kW	4.0 - 7.8 kW	4.8 - 9.3 kW	6.1 - 11.7 kW	6.6 - 12.7 kW
Max. usable input current (MPPT 1/MPPT 2)				18 A / 18 A		
Max. usable input current (MPPT 1+MPPT 2)				36 A		
Max. array short circuit current (1.5× Imax) (MPPT1/MPPT2)				27 A / 27 A		
Nominal input voltage		410 V	420 V	420 V	420 V	420 V
Operating voltage range				80 V - 600 V		
DC startup voltage				80 V		
MPP Voltage Range		200-480 V	200-480 V	240-480 V	250-480 V	270-480 V
Max. input voltage				600 V (1000 V optional)		
Admissible conductor size DC		AWG 14 - AWG 6 copper (solid / stranded)/ fine stranded)(AWG 10 copper or AWG 8 aluminium for overcurrent protective devices up to 60A, from 61 to 100A minimum AWG 8 for copper or AWG 6 aluminium has to be used), AWG 6 - AWG 2 copper (solid/stranded) MultiContact Wiringable with AWG 12				
Number of MPPT				2		
OUTPUT DATA		PRIMO 3.8-1	PRIMO 5.0-1	PRIMO 6.0-1	PRIMO 7.6-1	PRIMO 8.2-1
Max. output power	208 V/240 V	3800 VA/3800 VA	5000 VA/5000 VA	6000 VA/6000 VA	7600 VA/7600 VA	7900 VA/8200 VA
Output configuration				208/240 V		
Frequency range (adjustable)				45.0 - 55.0 Hz / 50 - 66 Hz		
Operating frequency range default for CAL setups				-/ 58.5 - 60.5 Hz		
Operating frequency range default for HI setups				-/ 57.0 - 63.0 Hz		
Nominal operating frequency				60 Hz		
Admissible conductor size AC				AWG 14 - AWG 6		
Total harmonic distortion				< 5.0 %		
Power factor range				0.85-1 ind/cap		
Max. continuous output current	208 V	18.3 A	24.0 A	28.8 A	36.5 A	38.0 A
	240 V	15.8 A	20.8 A	25.0 A	31.7 A	34.2 A
OCPD/AC breaker size	208V	25 A	30 A	40 A	50 A	50 A
	240 V	20 A	30 A	35 A	40 A	45 A
Max. Efficiency		96.7 %	96.9 %	96.9 %	96.9 %	97.0 %
CEC Efficiency		95.0 %	95.5 %	96.0 %	96.0 %	96.5 %

Let us divide this further into two parts.

1. Input data.
2. Output data.

There are a lot of details provided by the manufacturer in a datasheet but I am going to filter it out for you and focus on only those data points that will help us design the system in an efficient manner.

Input data:

TECHNICAL DATA FRONIUS PRIMO					
INPUT DATA	PRIMO 3.8-1	PRIMO 5.0-1	PRIMO 6.0-1	PRIMO 7.6-1	PRIMO 8.2-1
Recommended PV power (kWp)	3.0 - 6.0 kW	4.0 - 7.8 kW	4.8 - 9.3 kW	6.1 - 11.7 kW	6.6 - 12.7 kW
Max. usable input current (MPPT 1/MPPT 2)			18 A / 18 A		
Max. usable input current (MPPT 1+MPPT 2)			36 A		
Max. array short circuit current (MPPT 1/MPPT 2)			27 A / 27 A		
Nominal input voltage	410 V	420 V	420 V	420V	420 V
Operating voltage range			80 V - 600 V		
DC startup voltage			80 V		
MPP Voltage Range	200-480 V	200-480 V	240-480 V	250-480 V	270-480 V
Max. input voltage			600 V (1000 V optional)		
Admissible conductor size DC			AWG 14 - AWG 6 copper (solid / stranded / fine stranded) AWG 10 copper or AWG 8 aluminium for overcurrent protective devices up to 60A. From 61 to 100A minimum AWG 8 for copper or AWG 6 aluminium has to be used. AWG 6 - AWG 2 copper (solid / stranded) MultiContactWiringable with AWG 12		
Number of MPPT			2		

Maximum Permitted PV Power (kWp):

Recommended PV power (kWp)	3.0 - 6.0 kW	4.0 - 7.8 kW	4.8 - 9.3 kW	6.1 - 11.7 kW

This is the maximum input power that an inverter can take at its input. For a 3.8 kW inverter, the maximum power we can provide from solar panels to the input of this inverter is 6 kW.

    Our System size= 5 kW

    Maximum inverter input= 6 kW

So we have chosen the correct inverter. The maximum input power should never go above the solar capacity recommended in the datasheet. If the solar capacity is more than the inverter input capacity, we will need an inverter of higher power.

Number of MPPT:

Number of MPPT	2

This basically means there are two connection points in the inverter where the solar panels can be connected. We know that we need to connect 15 solar panels to the inverter.

It might seem like common sense to us that if there are two inputs to the inverter and 15 solar panels to be connected to it then let's just connect 7 panels to the first connection point (MPPT-1) and the remaining 8 to the second connection point (MPPT-2).

This will look something like this:

    MPPT-1: 8 panels

    MPPT-2: 7 panels

MPPT Voltage Range:

MPP Voltage Range	200-480 V	200-400 V	240-480 V	250-480 V

This is the voltage range that the inverter needs at its input for optimal performance. We need to connect our solar panels in such a manner so that the total voltage lies within the range of 250v to 480v. Let us take a quick look at the MPPT voltage of the solar panels.

Nominal Power Voltage - V_{MPP} (V)	38.1	38.3	38.5	38.7	38.9	39.1

The MPPT voltage of a single 350W solar panel is 38.9v!

Now you must be thinking,

INPUT TO INVERTER REQUIRED= 250v to 480v.

OUTPUT OF SOLAR PANELS= 38.9v.

So as per our previous inference let us see what happens if we connect 8 panels together.

8 panels together= 38.9 x 8= 311.2 v.

Similarly,

7 panels together= 38.9 x 7= 272.3 v.

And so the two combinations that we had chosen do lie within the MPPT voltage range.

RULE OF THUMB:

Choose the MPPT voltage to be in the center of the range.

What this means is that the ideal MPPT voltage should be:

$$\frac{\text{Low Range} + \text{High Range}}{2}$$

In this case =

$$\frac{200 + 400}{2} = 340 \text{ volts}$$

Our string voltage should ideally lie close to **340v**. The closer the better.

It is still okay if it is a bit far from that figure. The only thing which we should be worried about is that our string voltage should lie inside the MPPT range. Our string voltages of **311.2v** and **272.3v both lie inside the MPPT voltage range.**

Let us check the other parameters of MPPT in order to ensure that all the conditions are satisfied.

Open Circuit Voltage (Voc):

This is the voltage that you will get at the output of the solar panels when there is no device connected to the output of a solar panel. Let's take a quick look at the solar panel datasheet.

ELECTRICAL DATA @ STC	Product Code*: RECxxxTP2S 72					
Nominal Power - P_{MPP} (Wp)	330	335	340	345	350	355
Watt Class Sorting - (W)	-0/+5	-0/+5	-0/+5	-0/+5	-0/+5	-0/+5
Nominal Power Voltage - V_{MPP} (V)	38.1	38.3	38.5	38.7	**38.9**	39.1
Nominal Power Current - I_{MPP} (A)	8.67	8.75	8.84	8.92	9.00	9.09
Open Circuit Voltage - V_{OC} (V)	46.0	46.2	46.3	46.5	**46.7**	46.8
Short Circuit Current - I_{SC} (A)	9.44	9.52	9.58	9.64	9.72	9.78
Panel Efficiency (%)	16.5	16.7	16.9	17.2	17.4	17.7

Values at standard test conditions STC (airmass AM 1.5, irradiance 1000 W/m², cell temperature 77°F (25°C).
At low irradiance of 200 W/m² (AM 1.5 and cell temperature 77°F (25°C)) at least 95% of the STC module efficiency will be achieved.
* xxx indicates the nominal power class (P_{MPP}) at STC, and can be followed by the suffix XV for modules with a 1500 V maximum system rating.

It is evident from the panel datasheet, that there is a difference between Voc and Vmp.

When we don't connect an appliance to the output side of the panel the voltage comes out to be 46.7v.

Imagine a scenario when our 8 panels are connected to the inverter and the inverter is not supplying power. In this case, the Voc becomes:

= 46.7 x 8

= 373.6v

Let us see if this lies within the input range of the inverter.

TECHNICAL DATA FRONIUS PRIMO					
INPUT DATA	PRIMO 3.8-1	PRIMO 5.0-1	PRIMO 6.0-1	PRIMO 7.6-1	PRIMO 8.2-1
Recommended PV power (kWp)	3.0 - 6.0 kW	4.0 - 7.8 kW	4.8 - 9.3 kW	6.1 - 11.7 kW	6.6 - 12.7 kW
Max. usable input current (MPPT 1/MPPT 2)			18 A / 16 A		
Max. usable input current (MPPT 1+MPPT 2)			36 A		
Max. array short circuit current (1.5* Imax) (MPPT 1/MPPT 2)			27 A / 47 A		
Nominal input voltage	410 V	420 V	420 V	420V	420 V
Operating voltage range			80 V - 600 V		
DC startup voltage			80 V		
MPP Voltage Range	200-480 V	240-480 V	240-480 V	230-480 V	270-480 V
Max. input voltage			600 V (1000 V optional)		
Admissible conductor size DC	AWG 14 - AWG 6 copper (solid / stranded / fine stranded)AWG 10 copper or AWG 8 aluminium for overcurrent protective devices up to 60A, from 61 to 100A minimum AWG 6 for copper or AWG 4 aluminium to be used; , AWG 6 - AWG 2 copper(solid) stranded) MultiContact: Wiring able with AWG 12				
Number of MPPT			2		

The maximum voltage the inverter can take at its input is 600v. That means we can connect 8 panels. But wait, the Voc we took was from STC conditions and so we need to factor in the temperature as well.

Let's take the example of Michigan where the temperature measures from minimum -7.8 degrees Celsius to maximum 28.5 degrees Celsius. Let's jump to the second page of the panel datasheet again.

The effect of temperature on Voc is -0.30% per degree Celsius.

The Voc= 46.7v that we get is at 0 degree Celsius and so we need to evaluate it at the extreme points.

Let us evaluate the effect at the following two temperatures of Michigan.

Minimum temp= -26 degree Celsius

Maximum temp= 36 degree Celsius

If we had to calculate Voc at 1 degree the difference between STC and 1 degree is exactly 1, so our equation would look like this:

Voc (at 1 degree) = 46.7 − 0.3 % (46.7) = 46.55v

Therefore, the Voc at 1 degree will be **46.55v**, let us evaluate for Michigan.

$$\text{Voc (at -26 degree)} = 46.7 - [0.3\% (46.7)] \times (-26)$$
$$= \mathbf{50.34v}$$
$$\text{Voc (at 36 degree)} = 46.7 - [0.3\% (46.7)] \times (36)$$
$$= \mathbf{41.66v}$$

Let us calculate the Voc for 15 panels together at 5 degrees Celsius and 13 degrees Celsius.

$$\text{Voc (at -26 degrees for 8 panels)} = 50.34 \times 8$$
$$= 402.72v$$
$$\text{Voc (at -26 degrees for 7 panels)} = 50.34 \times 7$$
$$= 352.38v$$
$$\text{Voc (at 36 degrees for 8 panels)} = 41.66 \times 8$$
$$= 333.28v$$
$$\text{Voc (at 36 degrees for 7 panels)} = 41.66 \times 7$$
$$= 291.62v$$

The Voc at extreme temperatures also lies below the maximum input the inverter can take. Now we have ticked off two items from our checklist and only one final item remains before we can go ahead.

We also need to take care that the Voc at extreme conditions should not go below the lower voltage level of the MPPT range.

In our case the minimum voltage range is 240v. Hence, our sizing is correct, since even at extreme temperatures we reach 291.62v which is above 240v

Finally, we can now be sure that the selected combination is good to go!

Checklist:

1. DC/AC ratio should be around 1.3.

2. Check the MPPT range of the inverter.

3. The ideal combination of the solar panels should be such that the total Vmp of panels becomes equal to or close to the center of the MPPT range of the inverter.

4. Check the Voc of all the solar panels combined at the extreme temperatures of your area. It should always be below the maximum input voltage of the inverter and above the lower level of the MPPT range.

At this point you must be wondering, how do I actually connect the panels?

There are two wires at the output side of the solar panel and the inverter has two slots out of which we are going to use one.

How do I actually connect the panel wires to the inverter?

CHAPTER 17: CONNECTION

Until now we have dealt with the calculations from sizing, panel selection, and inverter selection. Finally, it's time to take hold of the wires and start connecting them together. Before we do that let's take a quick look at the types of connection methods.

Kenishirotie/Shutterstock.com

Our battery has two terminals, positive and negative. The flat side is the negative terminal and the other side is the positive terminal. Similarly, the solar panel can also be imagined as a battery, except instead of storing energy, it creates energy!

The red wire is the positive terminal.

The black wire is the negative terminal.

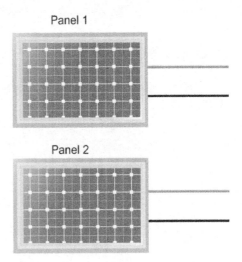

Let's say we are given a choice to connect the wires of panel 1 with panel 2. In how many ways can we do that?

1. Panel-1 **Red** to Panel-2 **Red**.

    Panel-1 **Black** to Panel-2 **Black**.

    OR

2. Panel-1 **Red** to Panel-2 **Black**.

Panel-1 **Black** to Panel-2 **Red**.

If we choose the first option, this is known as a **parallel connection**.

If we choose the second option, this is known as a **series connection**.

Parallel connection:

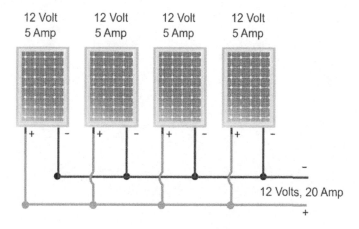

In Parallel System Current are Added and Voltage are Same

When the polarity of both the solar panels terminals that are to be connected is the same, it is known as a parallel connection. As you can see, the positive terminals of all the panels are grouped together. Similarly, the negative terminals of the panels are also grouped together.

In this case, when we connect the terminals in parallel, the output can be taken at any point. Remember, just like a battery, we need two terminals to function until we reach an inverter. If we want to connect the output of these four panels to the inverter we simply need to connect one red wire among all of them and similarly one black wire from any four.

But here's the problem. In electrical circuits, whenever we connect devices in parallel, the voltage does not increase. It remains the same through all panels and is also the same when combined. But we have seen in the previous chapters that we need Vmp to be at the center of the input range of the inverter. That was only possible when we combined the panels and each panel contributed its share to Vmp, that was how our optimum condition was achieved. But we cannot connect solar panels in parallel as Voc will be just 46.7v at STC and our ideal value was 340v. The closest we reached was 311.2v and to do that we needed to connect panels in series.

Series connection:

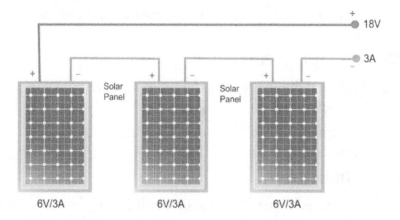

As you can see, the positive terminal of one panel is connected to the negative terminal of the other panel. Similarly, the positive terminal of the second panel is connected to the negative terminal of the third panel and so on.

CAUTION:

1. Never connect a positive terminal to the negative terminal of the same panel.
2. Never connect the positive terminal of the first panel to the negative terminal of the last panel.

We need to avoid the two conditions mentioned above. Once we have cautioned ourselves against these two conditions, we can move ahead in the designing process.

The last chapter told us that we need to connect 8 panels where the total voltage should add up to 311.2v and 7 panels where the total voltage is 272.3v at Vmp condition. We know that one solar panel at Vmp condition is capable of giving 38.9v, so we are left with one question in our mind. How to connect these two groups of panels.

Group-1: String 1 - 8 panels

Group-2: String 2 - 7 panels

But we don't know how to connect these 8 panels with each other (in series or parallel) or 7 panels with each other. Let us see the features of series and parallel connection and then conclude.

PARALLEL CONNECTION:
1. System voltage remains the same as individual panel voltage.
2. System current is the sum of the individual panel's current.

SERIES CONNECTION:
1. System voltage is the sum of the individual voltages of the panels.
2. Current remains the same throughout all the panels.

If you recall, we need the 8 panels to have a combined voltage of 311.2v. This is only possible in series type of connection as the voltage adds up in this case.

$$38.9 \times 8 = 311.2v$$

Similarly, for the other string of 7 modules:

$$38.9 \times 7 = 272.3v$$

We need to connect 8 panels in series separately and 7 panels in series separately.

Now, that we have connected 8 panels of string 1 in series, the total current of string 1 will be equal to the current of one panel.

So the Imp of one panel= 9 Amps.

ELECTRICAL DATA @ STC			Product Code*: RECxxxTP2S 72				
Nominal Power - P_{MPP} (Wp)	330	335	340	345	350	355
Watt Class Sorting - (W)	-0/+5	-0/+5	-0/+5	-0/+5	-0/+5	-0/+5
Nominal Power Voltage - V_{MPP} (V)	38.1	38.3	38.5	38.7	38.9	39.1
Nominal Power Current - I_{MPP} (A)	8.67	8.75	8.84	8.92	9.00	9.09
Open Circuit Voltage - V_{OC} (V)	46.0	46.2	46.3	46.5	46.7	46.8
Short Circuit Current - I_{SC} (A)	9.44	9.52	9.58	9.64	9.72	9.78
Panel Efficiency (%)	16.5	16.7	16.9	17.2	17.4	17.7

Values at standard test conditions STC (airmass AM 1.5, irradiance 1000 W/m², cell temperature 77°F (25°C).
At low irradiance of 200 W/m² (AM 1.5 and cell temperature 77°F (25°C)) at least 95% of the STC module efficiency will be achieved.
* xxx indicates the nominal power class (P_{MPP}) at STC, and can be followed by the suffix XV for modules with a 1500 V maximum system rating.

String= Collection (or group) of panels.

String 1: 8 modules in series,

    Total voltage= 311.2v (since voltage adds up in series).

    Total current= 9 A (Since current remains the same in series).

String 2: 7 modules in series.

    Total voltage= 272.3v (since voltage adds up in series).

    Total current= 9 A (Since current remains the same in series).

In this case we would have to connect string 1 to MPPT-1 and string 2 to MPPT-2.

Between the strings and inverter, we have to place a circuit breaker (next chapter).

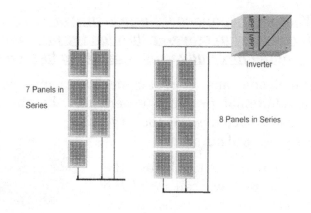

Understanding short circuit and fire hazard

You must have at least once heard this sentence in your life, "SHORT CIRCUIT LEADS TO FIRE". So, what is this short circuit? Why does it cause a fire hazard?

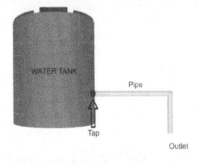

In this chapter, let us understand what exactly happens in a short circuit condition. Since we are going to deal with electrical circuits, it is my duty to educate you on one of the most dangerous situations in electrical engineering. I will also take you through the good practices to follow which will in turn safeguard us against this potential hazard. This practice will also be helpful for general house wiring purposes.

Let us continue with our water tank example. Imagine a water tank. Under normal conditions, the water will flow smoothly through the pipes. In these normal conditions there will also be some pressure that will be experienced in the pipeline. Since the pipes are durable enough, they can bear the pressure and function optimally.

<u>What would happen if the pressure increased suddenly? What if someone forces water through the pipe downwards towards the outlet forcefully using something like a motor?</u>

The obvious answer that comes to our mind is that there will be an additional pressure experienced by the pipes and if the pressure becomes way too large the pipes will burst!

Let's connect a tap before the pipe so that we can avoid the damage to the pipe. If we control the flow of water through the pipe, we can reduce the flow of water through the pipe and protect our pipe.

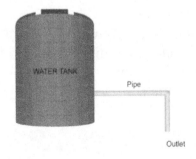

118

Now imagine there is another pipe which has no tap to control the flow of water.

Which tap will burst first?

Pipe 2 right?

If you look closely, we protected pipe 1 by placing an obstruction in its path i.e. a tap.

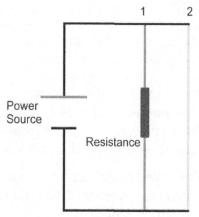

The purpose of this example was to explain the significance of a short circuit. It's obvious that any fluid will take the path of least resistance to reach its end point.

A similar thing happens when a huge amount of current flows through a conductor. If there are two conductors connecting two points, the current will choose the path of least resistance.

Let us consider the following two conductors:

Conductor 1 has some resistance.

Conductor 2 has no resistance.

Remember resistance is that part of an electrical circuit which opposes the flow of current.

Which conductor will current flow through?

The answer is conductor 2.

Just like the fluid flowing
through the path of least resistance, the current will also flow through the path of least resistance or the shortest part of the circuit, i.e. **short circuit!**

The current is lazy; it will always choose the path of least resistance.

At this point you might be wondering, why does a conductor catch fire in a short circuit condition?

Just like every pipe has a maximum pressure it can bear, every conductor has a maximum current it can withstand.

Every electrical conductor is rated for a certain current rating. You must have heard the wire dealer/ seller say 3 amps, 5 amps, 10 amps. These amps are the maximum amount of current that a conductor can bear.

Let's look at an example:

Now that we know the output of the inverter that can be used for powering our house, let us decide which wire we have to choose, let's take a quick look at the inverter datasheet.

OUTPUT DATA		PRIMO 3.8-1	PRIMO 5.0-1	PRIMO 6.0-1	PRIMO 7.6-1	PRIMO 8.2-1
Max. output power	208 V/240 V	3800 VA/3800 VA	5000 VA/5000 VA	6000 VA/6000 VA	7600 VA/7600 VA	7900 VA/8200 VA
Output configuration				208/240 V		
Frequency range (adjustable)				45.0 - 55.0 Hz / 50 - 66 Hz		
Operating frequency range default for CAL setups				~/ 58.5 - 60.5 Hz		
Operating frequency range default for HI setups				~/ 57.0 - 63.0 Hz		
Nominal operating frequency				60 Hz		
Admissible conductor size AC				AWG 14 - AWG 6		
Total harmonic distortion				< 5.0 %		
Power factor range				0.85 - 1 ind/cap		
Max. continuous output current	208 V	18.3 A	24.0 A	28.8 A	36.5 A	38.0 A
	240 V	15.8 A	20.8 A	25.0 A	31.7 A	34.2 A
OCPD/AC breaker size	208 V	25 A	30 A	40 A	50 A	50 A
	240 V	20 A	30 A	35 A	40 A	45 A
Max. Efficiency		96.2 %	96.9 %	98.9 %	96.9 %	97.0 %
CEC Efficiency		95.0 %	95.5 %	96.0 %	96.0 %	96.5 %

As we can see, the datasheet says the maximum output continuous current is 15.8A for a 240v system and 18.3A for a 208v system. For this example, I am considering 240v as most homes operate on this voltage.

The inverter datasheet tells us that an 18.3A current will flow when the inverter is loaded i.e. when appliances are connected at the output side.

What will happen if I use a wire which has a capacity of 18A?

The answer is, the wire will burn. The conductor starts heating up and catches fire. So we need to use a wire which can withstand more current. Our common sense might prompt us to use a wire which can withstand more than 18.3A leading us to select a wire of the next available size. But the problem with this approach is that we have not taken into account other factors such as temperature or derating of the conductor. We will be looking at all these factors in detail in the chapter on cable sizing and MSP (Main Service Panel) design.

There are two more steps remaining in this book after which you can finally go on your roof and start working.

CHAPTER 18: PROTECTION DEVICES

They say a good dog protects a house. Similarly, we electrical engineers say a good breaker protects the house and the rooftop system. In this chapter we will understand the importance of circuit breakers and other protection devices.

There are two main types of protection device:

MCB:

Imagine a scenario like this, a wire connecting the power source to the device (a bulb in this case). What will happen if suddenly there is a fault at the source? There is a sudden current spike, an increase in current due to a fault at the source end.

We discussed in the previous chapter that a high current can seriously damage the system. The wire can catch fire. Similarly, the bulb can get damaged too. So we need a device that can protect our bulb and also avoid the high current from leading to a short circuit and fire. Our intuition might tell us that there needs to be something in between the source and the appliance to protect our bulb.

Our intuition is correct once again, an MCB is placed between the source and the appliance. Let's redraw the diagram:

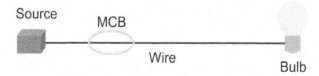

MCB stands for Miniature Circuit Breaker. This device is placed between the source and the appliance. Let us see how it operates:

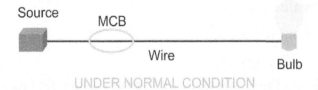

UNDER NORMAL CONDITION

Under normal conditions, the source is connected to the load as in the above diagram. But when the current goes above the rated value of the MCB, the MCB trips. The MCB disconnects the input and output. The circuit in this case looks like this:

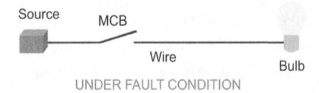

UNDER FAULT CONDITION

As you can see, the wire reaches the input of the MCB but the source and the appliance are not connected, since the MCB disconnects the source from the appliance.

At this point you might be wondering...

"Which MCB should I choose?"

Let us assume that the rated current of the appliances in total is 11 Amps. Generally, an MCB is connected for a group of devices, meaning your living room will have one MCB, similarly your bedroom will have one MCB. We do not connect an MCB to individual devices. The above example was just for understanding purposes.

Let's say the total rated current of your living room is 11 Amps. You can easily find the rated current on a label on your device. The best example of this is an LED bulb. The ratings are mentioned at the neck of the LED bulb. Similarly, add up all other current ratings from other devices. For now, we have assumed the total rated current to be 11 Amps.

RULE OF THUMB:

Multiply the rated current by 1.25, i.e. consider 25% extra of the rated value.

This comes out to be 11 x 1.25

$\qquad$ = 13.75 Amps.

Hence, we need to select a circuit breaker which will be of higher capacity than 13.75 Amps.

The MCB is available only in standard sizes of the following ratings:

1. 2 Amps
2. 6 Amps
3. 10 Amps
4. 12 Amps
5. 16 Amps
6. 20 Amps
7. 25 Amps
8. 30 Amps
9. 32 Amps.
10. 40 Amps.
11. 50 Amps.
12. 63 Amps.

The MCB selected in this case is **16 Amps.**

The MCB acts as an automatic switch. It connects the circuit in normal conditions and disconnects the circuit under fault conditions. We can also manually disconnect the MCB when we need to do so for repairing purposes.

Remember this rule every time you select a protection device.

The second type of protection used is a **FUSE**.

Fuse:

A Fuse is a thermo- electric element in a circuit which melts under high current and disconnects the input and output. The difference between a fuse and an MCB is that once a fuse undergoes a fault condition, it cannot be used again. Let's say a fault occurs today and the fuse successfully disconnects the input and output. Now, until you replace this fuse with a new one, your input and output side will be isolated from each other. But in the case of an MCB, resetting an MCB will again connect the input and output side and your circuit will be back in action.

Isolator:

An isolator is simply a switch of a higher capacity. Our switch board at home is generally 6 Amps. But what if the following situation occurs. Imagine your solar array is placed on the rooftop and the Main Service Panel is near your garage.

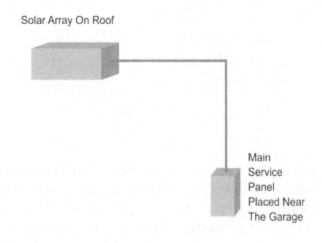

As we discussed earlier, the output of the inverter is connected to the Main Service Panel. Let us assume a scenario where there is a fault at the Main Service Panel end. Remember this is placed near the garage and this is connected to the inverter which is placed on the rooftop. How will you repair the MSP when the inverter is connected to the MSP?

Our school science lessons have trained us and made us believe that touching a live wire can lead to a fatal shock. So it becomes necessary to isolate or separate the inverter output and MSP.

Our intuition might tell us that it would be best to place an MCB in-between the inverter output and MSP.

While you can do that, there is one more device that can help us solve this problem. The device's name is an isolator.

The difference between an isolator and an MCB is that while an MCB will automatically trip under a fault condition, an isolator cannot automatically disconnect the circuit. Just like a switch, we have to manually rotate/ push the isolator and it will disconnect the inverter output and MSP.

These are the two protection devices that are most widely used for the protection of electrical system in rooftop solar. Our final block diagram including all the protection devices will look like this.

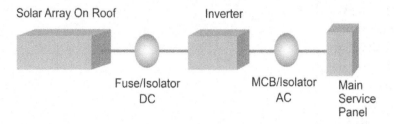

Just as we have isolated the MSP from the inverter by placing the protection device in between the inverter and the MSP, similarly we need to isolate the solar panel array and the inverter by placing the protection device in between them.

The following are three cases that can take place which will help us understand the importance of these devices.

Consider two of them to be isolators.

Case 1: Repair/ maintenance needed at the Solar Panel Array

Disconnect the isolator placed between solar panel array and the inverter.

Beware of live parts at the solar panel end. Since the solar panels will be generating energy, coming into contact with live parts might lead to electrocution.

Case 2: Repairs needed at the inverter

Simply disconnecting both the AC side and DC side isolators and inverter is fine. It has no live parts connected to it and can be repaired without threat.

Case 3: Repair/ maintenance needed at the Main Service Panel

Disconnect the isolator placed between the inverter and the Main Service Panel.

Beware of live parts at the Main Service Panel end. Since the Main Service Panel will also be connected to the utility power, coming into contact with live parts might lead to electrocution.

RULE OF THUMB:

Whether it be MCB, fuse, or isolator we always follow the same rule while sizing these protection devices. Remember to carefully take note of the input and output for a protection device. Let's take the example of a fuse placed in between a string and the inverter.

Just like we considered a 25% increase in the input side (or rated current) in our MCB example, we have to apply a similar rule when selecting a fuse.

We have to place a fuse between the string and the inverter. In this case the string current is the input and the inverter is the output.

The input current to the fuse= the total current of the string.

Total current of the string= 9 A. So we need to select a fuse:

    Fuse= 1.25 x 9 = 11.25A.

We will select a fuse of 12A (standard size).

Remember this rule every time you select a protection device.

CHAPTER 19: CONNECTING STRING TO INVERTER

We have two strings.

    First string - 8 panels in series.

    Second string - 7 panels in series.

We know that we need to connect the strings to the inverter and we need to place a fuse in-between. This is how our sequence should be:

    Strings - fuse/isolator - inverter.

This is how the string 1 connection should look once it is completed.

We would need two fuses. One for the positive terminal of the first module and one for the common negative. Similarly, we would also need two fuses for the second string.

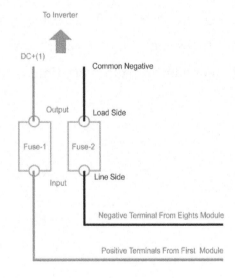

Let us take a look at how the two strings, i.e. string 1 and string 2 would look when they are combined.

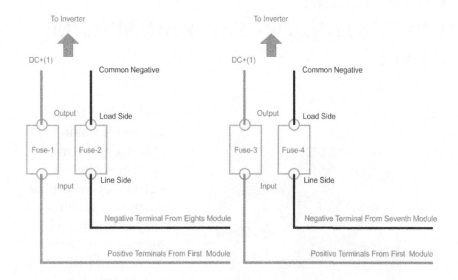

As we have seen in the previous chapter, the fuse selected for a string of 8 panels worked out to be **12 Amps.**

Similarly, in this case we have another string (string 2) of 7 panels and the total combined current of string 2 is also 9A (series connection).

So the fuse for string 2 will also be selected as **12 Amps**.

We will need 4 fuses of **12 Amps** as shown in the above diagram.

*Please note, some countries or local regulations may require you to have both fuse and isolator at the DC side, kindly check local regulations before designing a system.

CHAPTER 20: UNDERSTANDING THE MAIN SERVICE PANEL

Iconic Bestiary/Shutterstock.com

Remember, until now we've are talking about a grid connected system without a battery. So whatever we generate during sunlight hours will directly be connected to the grid (utility company). But we need a common point from where we can parallel or supply the power together for our house. We already have a metering point given to us by the utility company so we take this as the common point and then connect our rooftop solar system parallel to it.

At this point you might have this question:

"But how can utility power and solar rooftop power be connected together to supply power to my house?"

Remember energy and power work like filling up a bucket of water. You can fill a bucket of water by connecting it to multiple pipes. In this case we have two sources (pipes), one is solar and the other one is the existing power supplied by the utility company. We supply the power together.

Let us see what lies inside of an MSP.

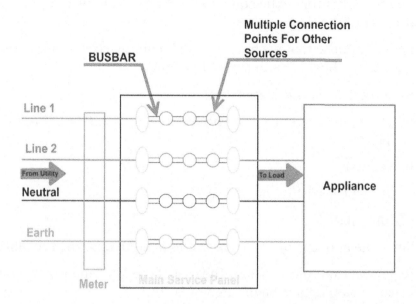

The above diagram depicts a single phase MSP/ two phase system.

Busbar: The busbar is like the common thread on the Main Service Panel. It is where you can connect the power coming from the utility company to solar power and/or other sources such as wind, or a diesel generator.

Connection points: These are the common points where we connect the incoming power from the inverter.

Meter: This is where all the power consumed from the utility supply gets noted and we are billed accordingly.

Just like a wire, the busbar also carries power.

At this point a question should concern us and that is...

Can I directly connect solar power to my busbars?

Well to answer that question we have to look at the busbar rating and also a special rule known as the 120% rule.

Step-1: Find your busbar rating, the busbar rating is generally given on the MSP. If it is not given on the MSP you have to assume the rating to be equal to the main incoming circuit breaker of the MSP.

Let us suppose that we have a main incoming circuit breaker of 50A, so the busbar rating would be 50A.

Now the 120 % rule states that:

Inverter max current + main breaker rating < 120 % of busbar rating.

In our case:

15.8 A +50A < 120% of 50A

65.8A < 60A

Since the above is not true we would need to upgrade the MSP.

Just as an example, consider the case if the MSP and main breaker would have been rated of 100A.

In such case the condition would have been:

65.8A < 120A

Since the above condition is satisfied, we do not need to upgrade.

To summarize, you can connect solar without an upgrade if:

Solar inverter output current + main breaker current < 120% of busbar's current.

Similarly, a three phase MSP will look like this:

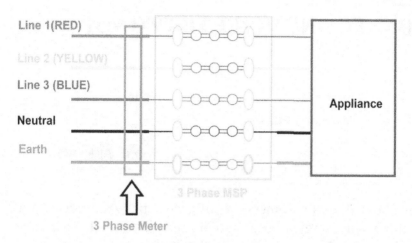

If your house has a three phase supply, you can compare the maximum output current from Fronius Symo to the MSP rating. Remember the only key difference is in the number of phase wires. The inverter maximum output current will always be same for each phase. Similarly, the rating of the MSP will also be the same for each phase. Now you just have to compare the rating from a Symo datasheet and compare it with your MSP rating as we have done in the example above.

CHAPTER 21:
CONNECTING INVERTER TO MSP

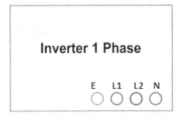
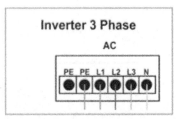

The above diagram depicts the difference between the output terminals of a single phase/two phase and three phase inverter. As I have previously mentioned, the difference lies in the additional phase. The neutral and earth terminals will be present. In case of a three phase system, two earth terminals might be provided for ease of connection.

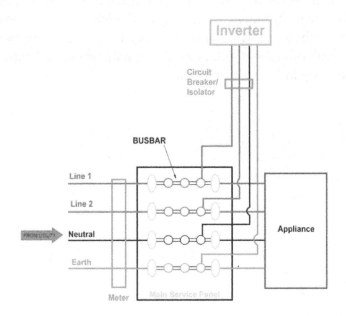

The above diagram depicts the connection diagram for a single phase system.

134

As we have discussed previously, the busbar is the common point of connection. We will connect the three output wires of our inverter in the following manner:

1. **Line-1:**

   The **L-1** of the inverter is connected to the **L-1** of the busbar at one of the common connection points.

2. **Line-2:**

   The **L-2** of the inverter is connected to the **L-2** of the busbar at one of the common connection points.

3. **Neutral:**

   The neutral of the inverter is connected to the neutral

   of the busbar at one of the common connection points.

4. **Earth:**

   The earth of the inverter is connected to the earth of the busbar at one of the common connection points.

Between these connections, we have our MCB Connection.

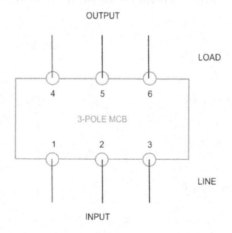

When we are connecting our inverter to our MSP,

   Inverter→ input.

   MSP→ output.

A 3-Pole MCB has 6 terminals,

3 serve as input and 3 serve as output.

On the MCB you will generally see mention of the input side and the output side.

You should see this type of symbol:

>Line== Input.

>Load==Output.

In such a case we will connect the inverter output to the lower side and the upper side will also have three wires coming out of it which will be connected to the MSP.

Remember, the phase connection should always be connected opposite a phase connection and a neutral connection should always be connected opposite a neutral connection.

Never cross connect the terminals

In our case, if I am connecting the phase of the inverter to terminal-1 and neutral to terminal- 2, I have to connect terminal-3 to phase of MSP and terminal-4 to neutral of MSP.

Terminal-1➔ Line-1 of inverter

Terminal-4➔ Line-1 of MSP.

Terminal-2➔ Line-2 of inverter.

Terminal-5➔ Line-2 of MSP.

Terminal-3➔ neutral of inverter.

Terminal-6➔ neutral of MSP.

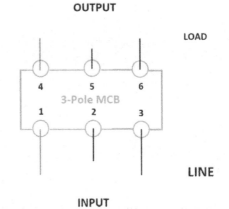

The terminals in the same vertical axis are internally connected. In our case, under normal conditions:

Terminal-1--- Connected to--- Terminal-3.

Terminal-2--- Connected to--- Terminal-5.

Terminal-3--- Connected to--- Terminal-6.

Never cross connect the terminals! It can lead to a short circuit and fire hazard!

Cautions while connecting MCB

1. Line is connected to the input side.
2. Load is connected to the output side.
3. Never connect input to load and output to line whenever line and load are mentioned on the MCB.
4. Do not cross connect the terminals.

There is also one more type of MCB available on the market which is bi-directional,

In this case you can ignore point 3 of the above checklist but all other points remain the same. The bi-directional MCB will have no mention of line and load terminals, it will not have any single mention of which side line or load is!

Here is an example:

As you can see, there is no mention of the line and load side.

In such circumstances, you can connect the input on the lower side or to the upper side.

Let's recap the difference.

The major difference between the two is:

Unidirectional:

You cannot connect inverter output to the upper side (load side) and MSP to the top side (line side) of the unidirectional MCB.

Bi-directional:

You can connect inverter output to the upper side (load side) and MSP to the top side (line side) of the bi-directional MCB.

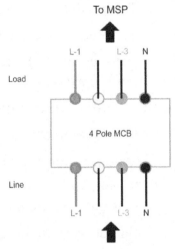

OR

You can connect inverter output to the lower side (line side) and MSP to the top side (load side) of the bi-directional MCB.

A three phase system also needs a circuit breaker. In this case we will be using a 4 pole circuit breaker.

A 4 pole circuit breaker consist of 3 single phases and one neutral phase.

A 4-pole circuit breaker will look like this:

G0d4ather/Shutterstock.com

MCB sizing:

Going by our rule of thumb of 25% more than the input current, we will select an AC MCB as follows:

AC MCB sizing:

MCB rating= 1.25 x maximum inverter output current.

In the case of a 3.8kw inverter:

OUTPUT DATA		PRIMO 3.8-1	PRIMO 5.0-1	PRIMO 6.0-1	PRIMO 7.6-1	PRIMO 8.2-1
Max. output power	208 V/240 V	3800 VA/3800 VA	5000 VA/5000 VA	6000 VA/6000 VA	7600 VA/7600 VA	7900 VA/8200 VA
Output configuration				208/240 V		
Frequency range (adjustable)				43.0 - 55.0 Hz / 50 - 66 Hz		
Operating frequency range default for CAL setups				/ 58.5 - 60.5 Hz		
Operating frequency range default for HI setups				-/ 57.0 - 63.0 Hz		
Nominal operating frequency				60 Hz		
Admissible conductor size AC				AWG 14 - AWG 6		
Total harmonic distortion				< 5.0 %		
Power factor range				0.85 1 ind./cap		
Max. continuous output current	208 V	18.3 A	24.0 A	28.8 A	36.5 A	38.0 A
	240 V	15.8 A	20.8 A	25.0 A	31.7 A	34.2 A
OCPD/AC breaker size	208 V	25 A	30 A	40 A	50 A	50 A
	240 V	20 A	30 A	35 A	40 A	45 A
Max. Efficiency		96.7 %	96.9 %	96.8 %	96.9 %	97.0 %
CEC Efficiency		95.0 %	95.5 %	96.0 %	96.0 %	96.5 %

Maximum inverter output current= 15.8A.

AC MCB rating= 1.25 x 15.8.

$\quad$ = 19.75A.

So we will select a **20A AC MCB.**

CHAPTER 22: CABLE SIZING

We know that the solar panels will be connected to the inverter. But we need wires (cables) to connect them. The wire you choose should be 10AWG PV copper wire. Do not use wires below that rating as that is a standard used worldwide. 10 AWG PV wires are special wires made for solar PV applications. These wires possess certain special properties such as UV resistance and added layers of insulation.

The second step is to connect the inverter to the MSP. In this case we need to refer to standard codes mentioned by electrical authorities.

Progressing with our previous example, the maximum inverter output current is 15.8A.

The minimum ampacity the wire should have= 1.25 x 15.8= 19.75A.

Until now we have been talking in terms of three wires (phase, neutral, and earth) in single phase. But we do not separately carry three different wires, we bundle it together, and call it a 3-core cable.

The cable that I would recommend is 3-core for a single phase system, 4-core for a two-phase system and 5-Core for a three phase system. There are two ways to select the cable. One would be to select the cable based on the standards followed. The Regulatory Authority makes electrical standards for this purpose. European countries follow IEC standards, and the USA follows NEC standards. Hence, it is best to go searching for these standards online and refer to them.

Steps to use for cable sizing:

1. Identify the maximum current (15.8A in our case- maximum inverter output).

2. Take overcurrent protection into account (15.8 x 1.25= 19.75A).
3. Search for a wire from table 310.15.B (16)[1] of capacity higher than 19.75A in the 90 degree Celsius column(as highlighted in black) as the wire we would be using is THWN-2.
4. We can observe that in the third column we have 25A(as highlighted in red).
5. This is the wire we can select. The wire to be selected in this case will be 14AWG.

NOTE: Generally speaking, if you satisfy the above criteria and your cable lengths are in the range of 10- 15 meters, your cable sizing will usually be correct, however readers are advised to verify their calculations as per NEC requirements. The detailed NEC calculations are out of the scope of this book and there are already many documents and online calculators available.

---

[1] Source for table: https://www.tooltexas.org/wp-content/uploads/2018/08/2017-NEC-Code-2.pdf

Table 310.15(B)(16) (formerly Table 310.16) Allowable Ampacities of Insulated Conductors Rated Up to and Including 2000 Volts, 60°C Through 90°C (140°F Through 194°F), Not More Than Three Current-Carrying Conductors in Raceway, Cable, or Earth (Directly Buried), Based on Ambient Temperature of 30°C (86°F)*

Size AWG or kcmil	Temperature Rating of Conductor [See Table 310.104(A).]						Size AWG or kcmil
	60°C (140°F)	75°C (167°F)	90°C (194°F)	60°C (140°F)	75°C (167°F)	90°C (194°F)	
	Types TW, UF	Types RHW, THHW, THW, THWN, XHHW, USE, ZW	Types TBS, SA, SIS, FEP, FEPB, MI, RHH, RHW-2, THHN, THHW, THW-2, THWN-2, USE-2, XHH, XHHW, XHHW-2, ZW-2	Types TW, UF	Types RHW, THHW, THW, THWN, XHHW, USE	Types TBS, SA, SIS, THHN, THHW, THW-2, THWN-2, RHH, RHW-2, USE-2, XHH, XHHW, XHHW-2, ZW-2	
	COPPER			ALUMINUM OR COPPER-CLAD ALUMINUM			
18**	—	—	14	—	—	—	—
16**	—	—	18	—	—	—	—
14**	15	20	25	—	—	—	—
12**	20	25	30	15	20	25	12**
10**	30	35	40	25	30	35	10**
8	40	50	55	35	40	45	8
6	55	65	75	40	50	55	6
4	70	85	95	55	65	75	4
3	85	100	115	65	75	85	3
2	95	115	130	75	90	100	2
1	110	130	145	85	100	115	1
1/0	125	150	170	100	120	135	1/0
2/0	145	175	195	115	135	150	2/0
3/0	165	200	225	130	155	175	3/0
4/0	195	230	260	150	180	205	4/0
250	215	255	290	170	205	230	250
300	240	285	320	195	230	260	300
350	260	310	350	210	250	280	350
400	280	335	380	225	270	305	400
500	320	380	430	260	310	350	500
600	350	420	475	285	340	385	600
700	385	460	520	315	375	425	700
750	400	475	535	320	385	435	750
800	410	490	555	330	395	445	800
900	435	520	585	355	425	480	900
1000	455	545	615	375	445	500	1000
1250	495	590	665	405	485	545	1250
1500	525	625	705	435	520	585	1500
1750	545	650	735	455	545	615	1750
2000	555	665	750	470	560	630	2000

*Refer to 310.15(B)(2) for the ampacity correction factors where the ambient temperature is other than 30°C (86°F). Refer to 310.15(B)(3)(a) for more than three current-carrying conductors.
**Refer to 240.4(D) for conductor overcurrent protection limitations.

CHAPTER 23: RACKING IT UP

During the course of this book we have focused only on the electrical aspect of the project. There is however, an important part that remains, and that is the "**mechanical aspect**" of this project.

The solar panels used need to be mounted on the roof in a concrete manner. They should be able to withstand the wind and support themselves. So just like we fit our furniture with screws and other fitting equipment, solar panels also can be fixed in a similar manner.

There are two layouts in which we can place the panels on the roof.

1. Portrait

In the portrait arrangement, the shorter side is parallel to the ridge of the roof.

Depending on the area available on the roof and obstruction on your roof, the arrangement may have to be selected.

In case the panels do not fit on the roof, then you can consider another type of arrangement i.e. landscape.

2. Landscape

In this type of arrangement, the longer side of the panel is parallel to the ridge of the roof.

Depending on the type of your roof, you can choose the type of arrangement you want.

Sometimes a combination of portrait and landscape arrangement is used.

Let us now move onto calculating the number of screws, rails and other accessories required to mount the solar panels.

Rails:

These are the rails on which our solar panels will be mounted. They give uniformity to the structure along with support.

Elena Elisseeva/Shutterstock.com

Rafters:

The rails are attached to the rafters of the roof. The rafters are slabs running perpendicular to the ridge of the roof. The rafter spacing is generally 2 ft. apart.

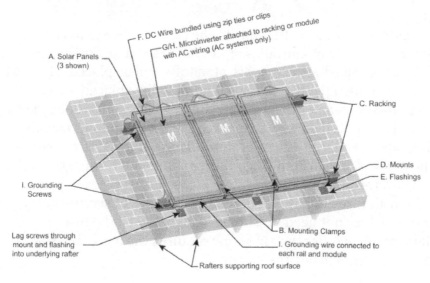

Standoff:

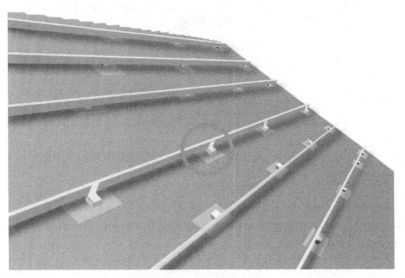

The standoff connects the rails to the rafters. The stand offs are generally placed in multiples of the rafter spacing. For example, if the rafter spacing is 2ft apart, standoffs will be placed 4ft apart from each other. Or if a second standoff will be placed 4ft apart from the first standoff, thereafter the third standoff will be placed at 8ft,12ft, and so on with respect to the first standoff. Once the standoffs are placed, the rail can be firmly attached to the roof. The process is to drill through the standoff and through the rail and also through the rafter. After this we insert the screw and tighten all three together. There are two types of screw. One type of screw does not need nuts to be fixed between the three elements. The other type of screw requires the nut to be fixed to the screw from inside the house.

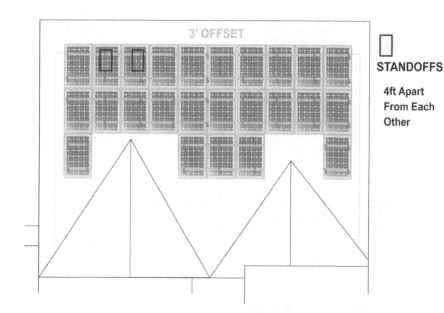

Mid clamp:

As the name suggests, the mid clamp is placed in-between two panels. The mid clamp holds the two adjacent panels together.

Sornthna/Shutterstock.com

Wichien Tepsuttinun/Shutterstock.com

End clamp:

Similar to the mid clamp, the end clamp holds the panels together. As the name suggests, the end clamp is placed at the two extreme ends of the arrangement.

Imagine you have a solar panel array like this. There are 12 panels in total, comprising of 6 in each row. We will need 8 end clamps which will be placed at each end.

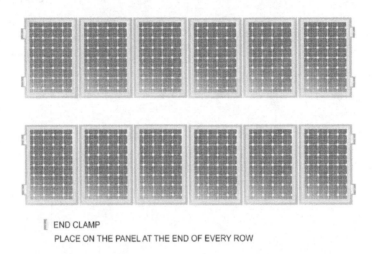

END CLAMP
PLACE ON THE PANEL AT THE END OF EVERY ROW

Finally, our arrangement will look like this:

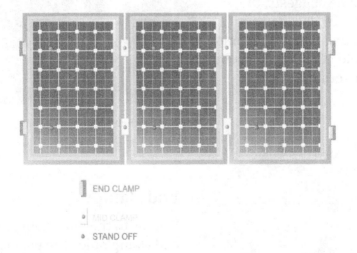

END CLAMP
MID CLAMP
STAND OFF

The racking arrangement is also dependent on the type of roof.

1. Type of roof:

The type of roof can be of any of the following categories:

148

- Flat concrete roof
- Pitched metal roof
- Pitched tile roof- S type
- Pitched tile roof- W type
- Pitched tile roof- flat type

2. System size:

Before buying the racking arrangement, we need to know the exact configuration of our modules. In our example we have selected 15 modules of 350 Watts each.

As with the panel and the inverters, we have two options to buy the racking and other accessories.

3. Offline:

In case you have a distributor/dealer of the racking system nearby you and you wish to opt for this option, you have to take the following thing with you:

- Photograph of your roof
- Panel configuration

4. Online:

In case you wish to order the racking arrangement online, first draw a tentative hand sketch or a diagram in a graphics package as to how your solar panel array will look on your rooftop. We have already calculated the dimensions of our roof, so keeping those dimensions in mind, let us prepare a rough sketch.

Steps to draw the rough diagram in a graphics package

Draw the roof as a rectangle using the length and width calculated during the previous chapters, the length and width calculated should exclude the fire safety setback. Let us assume the dimensions of the roof to be 10m x 7m (32.8 ft. x 22.96 ft.) excluding the fire safety setback. Just draw a random rectangle which would look something like this:

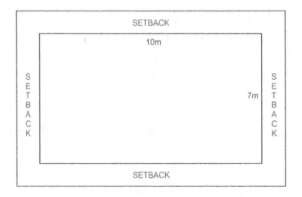

Remember, these are not to scale but are more our assumption, since to make them to actual scale, we would need the help of professional software such as AutoCAD. But for now, we can do it without them!

Now that we need 15 panels on our roof, our intuition might prompt us into saying;

- 8 panels in the first row
- 7 panels in the second row

And yet again, our intuition is correct; we know that the dimensions of each panel are ~2m x 1m. (6.56 ft. x 3.28 ft.).

If we opt for a portrait arrangement, our arrangement might look like this:

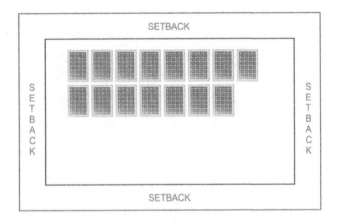

In this case I have assumed a pitched roof so I have not kept any interrow spacing. A rough diagram like this should give us an idea before we start ordering racking online.

With this roof I have assumed that there are no obstructions on the roof. So we can freely place our panels in two rows. The racking calculations in this case are easy.

- 4 rails (2 for each row)

When you are counting the number of rails, also approximately estimate the rail length. For example, in this case, we have 8 modules in the first row and the horizontal dimension of each row is ~8m. Keep this in mind as it will be helpful while choosing rails to buy.

- 8 end clamps
- 26 mid clamps (1 between two panels)

We will be calculating the exact number of standoffs and screws on the website from where we are ordering the racking.

But what if there are vents and other obstructions on the roof? The 15 Panels can be placed in the following manner,

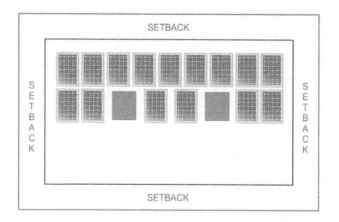

As you can see, the arrangement has changed due to the obstructions in the second row. But we have still managed to place the panels. In this case, we would need:

- 8 separate rails
- 16 end clamps
- 22 mid clamps

The purpose of this example is to give you an insight regarding how to have a generic idea of how our rooftop solar system will look.

You can similarly try out the landscape arrangement and see the results for yourself.

CHAPTER 24: RACKING WEBSITES

There are two ways to order a racking system online. The first way is where you simply give your system size, number of modules, and arrangement type, and the sales team send you a quote for the same. This is just like going to a physical shop and telling them your needs.

The second method is a bit more complex. In this method, we actually draw and configure the system on the seller's website. Let us take the example of Renusol.

Open your web browser and type in:

https://www.renusol.com/en/

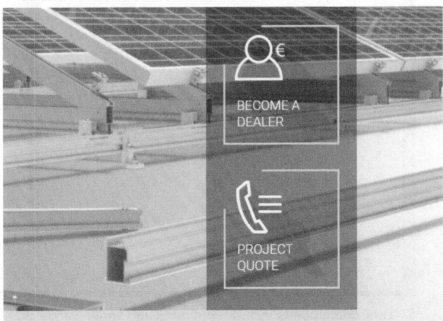

You can send in your requirements to them and get a quote. But if you want to do it yourself, go to the bottom of the website and click on PV-Configurator 3.0:

https://www.pv-configurator.com/login

You need to register first and then login to try the configurator.

They also have a detailed video dedicated to how to configure your array on their software.

It works like this:

After you have logged into their site, you need to fill in the following information step-by-step:

1. Project name, design number.
2. Enter the location of your house.
3. Enter the wind load. This can be selected from the database where you have to select your area.
4. Enter the snow load. This can be selected from the database where you have to select your area.

5. Other data points include roof pitch and panel warranty.

6. After you are done filling in all the details, you can then draw the array on their website just like we have drawn it in Paint.

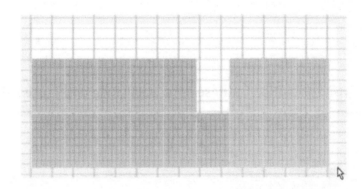

7. Finally you will have a detailed bill of materials which will consist of rails, clamps, and other mounting accessories.

Renusol has a detailed video on how to configure your system on the following link:

Video Link: https://youtu.be/jnxVchrPGQk

You can configure your rooftop array on another website:

The name of the website is SnapNRack: https://snapnrack.com/

Enter the website and click on configurator-Ultra Rail.

Here also, you need to enter the details such as panel type, wind speed, etc.

After you have filled in these details, click on configure sub array.

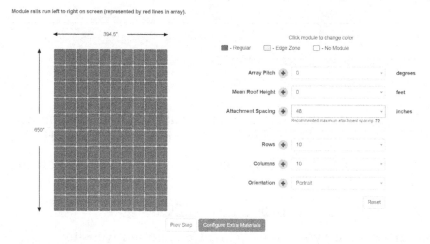

Now, you will have to design the array by adding/ subtracting rows/ columns.

Attachment spacing depends on the rafter spacing of your house. In our case, the rafter spacing is 2 ft. so our standoff spacing will be 4ft apart and the attachment spacing will be 48 inches.

After you are done with this, click on configure extra materials.

Preliminary BOM:

Part Number	Description	Qty	Extra
232-02450	SNAPNRACK, UR-40 RAIL, 168IN, SILVER	50	
242-01213	SNAPNRACK, UR-40 SPLICE, SILVER	40	
242-02052	SNAPNRACK, BONDING MID CLAMP, 38-51MM (1.49-2.00IN), SILVER	180	
242-02215	SNAPNRACK, UNIVERSAL END CLAMP	40	
232-02452	SNAPNRACK, UR-40 END CAP	40	
242-01220	SNAPNRACK, ULTRA RAIL UMBRELLA L FOOT, BLACK	180	
242-02256	SNAPNRACK, UMBRELLA LAG, TYPE 3, 4IN, SS	180	
232-01375	SNAPNRACK, COMP FLASHING, 9IN X 12IN, BLACK ALUM	180	
242-02101	SNAPNRACK, GROUND LUG ASSEMBLY, 6-12 AWG	10	

You will then be given a choice to add extra materials if any. Generally, this is an optimum design so there will be no need for any extra materials.

After this, click on configure bill of materials. There you have it, you are done with designing the racking for your PV rooftop system.

Here are a few other websites:

K2 Mounting Systems: https://k2-systems.com/en/start

Schletter Mounting Systems: https://www.schletter-group.com/

Play around on these websites, compare the quotes and decide accordingly.

CHAPTER 25: THE FINAL CHECKLIST

Let us recap all the steps from start to finish that we need to complete in order to successfully install a rooftop solar PV system.

Step 1: Calculate the size of the system required.

Skip to page 2 of your electricity bill,

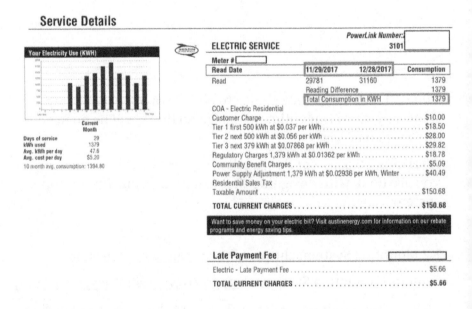

This is a 33 day bill.

Number of consumed units/ number of days =   1379
                                              ―――
                                               33

Number of consumed units/ number of peak days = 41 units per day

RULE OF THUMB: 1 Watt produces 0.0045 units per day (Varies with size, check the irradiance map)

System size required = $\dfrac{41}{0.0045}$ = 9,111 watts = 9 kW

So our system size is 9 kW.

Step 2: Roof assessment

Locate your country with regards to equator position and find the best side of the roof for solar.

1 kW requires 6-8 sq. m (64-86 sq. ft.) area. So our system will require 72 sq. m (775 sq. ft.)

Now go ahead and calculate the area of your roof. Let us call it (A)

Now subtract the area of obstruction from it. Let the area of obstruction be (R)

The remaining area= (A- R)

Now subtract the area of fire safety setback from each side. Let the area for fire safety setback be FS.

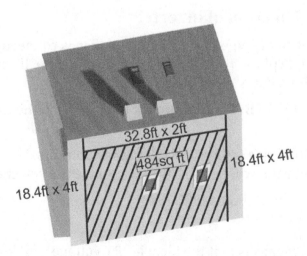

We get a useful area of 45sq m (484sq ft.). So we can a fit a maximum of 5kW. Now let us evaluate the ideal roof orientation.

If you live in a country above the equator you can place the panels on the following roof faces:

1. South (first preference).
2. West (second preference).
3. East (third preference).

And, if you live in a country below the equator you can place the panels on the following roof faces:

1. North (First preference).
2. West (second preference).
3. East (third preference).

If the total area available is less than the total area required adjust the wattages of the panel so as to fit the maximum kW on the roof.

Step 3: Selection of solar panel brand

Since the solar panels are going to be powering your house for the next 20-25 years make sure you choose a reliable market brand. Read the brochure/ datasheet for warranty types and online reviews.

Step 4: Selection of inverter

First identify the type of system your house is connected to. Either single phase (or two phase) or three phase. If single phase select Primo, if it is three phase select Symo.

RULE OF THUMB: DC/AC ratio should lie within the range 1.25-1.35.

If our system size is 5kW and we are selecting a 3.8kW inverter, the DC/AC ratio will be= 5/3.8= 1.31. So we can select this size.

Go to the center of the MPPT range;

What this means is that the ideal MPPT voltage should be:

$$\frac{\text{Low Range} + \text{High Range}}{2}$$

$$\frac{200 + 400}{2} = 340 \text{ volts}$$

Check for the Voc of the panels when connected together in series. Apply the percentage effect due to temperature on Voc. Check both the extreme temperatures of your area. The combined Voc should not go below the minimum input of inverter and the combined Voc should not go above the maximum input voltage to the inverter.

Step 5: Check for MSP upgrade

The rating of the busbar inside the MSP will be written on the metallic panel of the MSP.

Apply the 120% rule to check whether an MSP upgrade is required or not.

If an MSP upgrade is required, you need to contact your local electrician. Once the upgrade is done, we can then proceed ahead with the other steps of installation.

Step 6: Circuit breaker selection

We have to choose two circuit breakers/ isolators. One will be placed between the solar panels and the inverter. This will be a DC circuit breaker. The other circuit breaker will be placed between the inverter and the MSP. This will be an AC circuit breaker.

The rating of the fuse= 1.25 x total output current of the string.

The rating of the AC circuit breaker/ isolator= 1.25 x max output current of the inverter.

Step 7: Cable sizing

Select the wire to connect the PV array to the inverter as 4 sq. mm. (10 AWG).

Now check the maximum output current of the inverter and multiply it by 1.25 (25% more).

Select the wire from the table 310.15.B(16).

NOTE: Generally speaking, if you satisfy the above criteria and your cable lengths are in the range of 10- 15 meters, most of the time your cable sizing will be correct, however readers are advised to verify their calculations as per NEC requirements. The detailed NEC calculations are out of the scope of this book and there are already many documents and online calculators available.

Step 8: Estimate the mechanical requirements

Draw a rough sketch based on the available area as to how the solar panels will look once they are placed on the roof. Once that is done, go ahead and use the configurator to actually implement the design and get a quote/ bill of material from the website.

Step 9: Buy the items

Whether you choose to physically go into a shop and buy everything or order online, double check your calculations to optimize the cost before buying.

CHAPTER 26: INSTALLATION

In the beginning of this book I promised you that I would be there with you on this journey to your rooftop. Finally, it's time to go on the roof and put everything into action. All the previous chapters were dedicated to design and learning about how each component works. In this chapter, we be implementing everything.

Before proceeding, a word of caution:

Working on the roof has associated risks such as falling/sliding off the roof and other dangers that can result in serious injury. So the reader is advised to use his own judgment to decide if he/she wants to go ahead with the installation. It is advised to hire a professional to get the job done. Since, we will be dealing with onsite work dealing with high voltages and heavy mechanical equipment; it is better to always safeguard ourselves from potential threats.

Mechanical mounting

Before we start dealing with the mechanical installation part, here are a few things you need to be careful of:

PRECAUTIONS:

1. Always wear a safety harness.
2. Always wear a safety helmet.
3. Other PPE (personal protective equipment) as specified by your local regulatory authority.

STEP 1: Fix the standoff using the known measurements of your roof rafters.

STEP 2: Mount the rail on the stand offs.

STEP 3: Start placing the panels and mid clamps in between them.

STEP 4: At the end of each row place the end clamps.

In addition to these steps it would be advisable to read through the installation manual of the racking supplier since the methods may vary based on the location/ roof type.

Electrical connection

Before we start dealing with the electrical connection part, here are a few things you need to be careful of:

PRECAUTIONS:

1. Always wear a safety harness.
2. Always wear electrical safety gloves.
3. Do not touch any bare/ uninsulated wire.
4. Always switch off the main service panel before doing any electrical work.
5. Other PPE (personal protective equipment) as specified your local regulatory authority.
6. It is wise to take a professional's advice.

In our example we are connecting 15 modules in series.

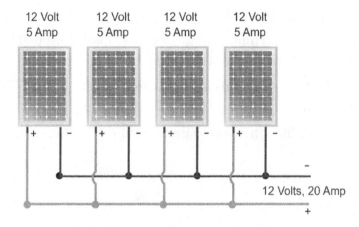

In Parallel System Current are Added and Voltage are Same

1. As in the above image where we have connected 4 modules in series, similarly, connect all the 8 modules in series in the same manner by connecting the positive of the first module to the negative of the second module, positive of the second module to the negative of the third module and so on until the last module.

Do not connect the positive terminal of the first module to the negative terminal of the last module (8th in this case), as it will result in a short circuit.

2. Connect the DC MCB between the string and inverter.

This will be a DC MCB. Check for the load side and line side of the MCB. First switch off the MCB. The positive terminal of the first module and the negative terminal of the 8th module will be connected to the line side of the MCB in the two separate ports given at the line side. The output (load side of the MCB) goes to the inverter input. Repeat the same for string two.

3. Connect the fuse to the inverter.

We will need two DC MCBs. One for each string.

The inverter will have two MPPTs, it will be labeled as follows:

MPPT-1: DC+ (1)

MPPT-2: DC+ (2)

The two output terminals of the MCB- 1 (load side) will be connected to the inverter (MPPT-1).

The two output terminals of the MCB- 2 (load side) will be connected to the inverter (MPPT-2).

The negative terminal would be common.

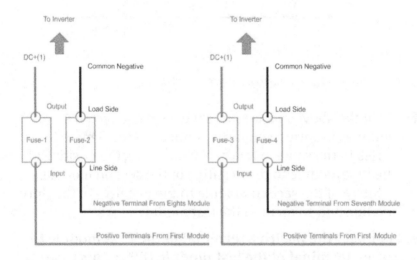

String 1 will have two outputs:

(+) positive terminal from first module.

(-) negative terminal from eighth module.

Positive from first module ⮕ fuse-1 ⮕ MPPT (+1) of the inverter.

Negative from eighth module ⮕ fuse-2 ⮕ Common negative of the inverter.

Similarly,

String 2 will have two outputs:

(+) positive terminal from first module.

(-) negative terminal from seventh module.

Positive from first module⟶ fuse-3⟶MPPT (+2) of the inverter.

Negative from seventh module⟶fuse-3⟶Common negative of the inverter.

Total string current= 9Amps.

Hence, fuse rating= 1.25 x 9 = 11.25Amps.

Therefore, we will select four fuses, one for each wire of the rating **12Amps (standard size)**.

4. Connect the inverter to AC MCB.

The inverter will have three output terminals.

Line, neutral and earth.

Switch off the MCB before making the other connections.

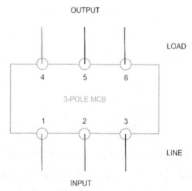

You can consider L and N to be the same as the positive and negative coming from the solar panels and connect the MCB accordingly.

AC MCB Sizing

MCB Rating= 1.25 x maximum inverter output current.

In the case of a 3.8kW inverter;

OUTPUT DATA		PRIMO 3.8-1	PRIMO 5.0-1	PRIMO 6.0-1	PRIMO 7.6-1	PRIMO 8.2-1
Max. output power	208 V/240 V	3800 VA/3800 VA	5000 VA/5000 VA	6000 VA/6000 VA	7600 VA/7600 VA	7900 VA/8200 VA
Output configuration				208/240 V		
Frequency range (adjustable)				45.0 - 55.0 Hz / 50 - 66 Hz		
Operating frequency range default for CAL setups				59.3 - 60.5 Hz		
Operating frequency range default for HI setups				57.0 - 63.0 Hz		
Nominal operating frequency				60 Hz		
Admissible conductor size AC				AWG 14 - AWG 6		
Total harmonic distortion				< 5.0 %		
Power factor range				0.85-1 ind/cap		
Max. continuous output current	208 V	18.3 A	24.0 A	28.8 A	36.5 A	38.0 A
	240 V	15.8 A	20.8 A	25.0 A	31.7 A	34.2 A
OCPD/AC breaker size	208 V	25 A	30 A	40 A	50 A	50 A
	240 V	20 A	30 A	35 A	40 A	45 A
Max. Efficiency		96.7 %	96.9 %	96.9 %	96.9 %	97.0 %
CEC Efficiency		95.0 %	95.5 %	96.0 %	96.0 %	96.5 %

Maximum inverter output current= 15.8A.

AC MCB rating = 1.25 x 15.8

= 19.75A.

So we will select a **20A AC MCB**.

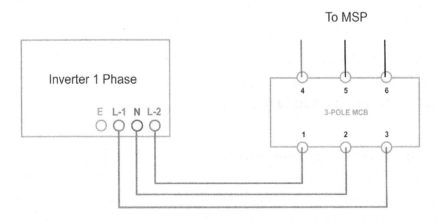

5. Connect the AC MCB to the MSP

Before you commence with this step, be sure to switch off the MSP. This can be done by switching off the isolator/isolators and/or circuit breaker/ circuit breakers at the input side of the MSP.

Once you have successfully checked whether your MSP needs an upgrade or not and dealt with it, the next step is to connect the AC MCB to MSP.

The line terminal coming out from the AC MCB should be connected to the line port of the busbar of your MSP.

The neutral terminal coming out from the AC MCB should be connected to the neutral port of the busbar of your MSP.

The earth terminal coming out from the AC MCB should be connected to the earth port of the busbar of your MSP.

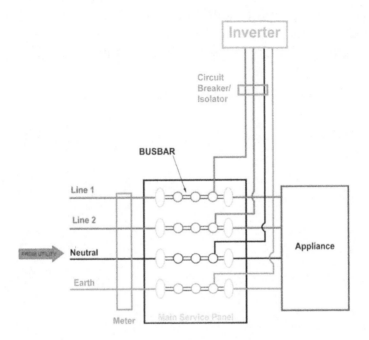

That's it!

Now go ahead and switch on all the MCBs.

Your PV rooftop project is ready!

"But how do I know if my solar array is generating power?"

There will be an LCD screen on the inverter which will be showing the amount of power generated by your PV system. You can see those numbers and verify.

Cable sizing:

Select the wire to connect the PV Array to the inverter as 4 sq. mm. (10 AWG).

Now check the maximum output current of the inverter and multiply it by 1.25 (25% more).

Select the wire from the table 310.15.B(16) .

NOTE: Generally speaking, if you satisfy the above criteria and your cable lengths are in the range of 10- 15 meters. Most of the time your cable sizing will be correct, however readers are advised to verify their calculations as per NEC requirements. The detailed NEC calculations are out of the scope of this book and there are already many documents and online calculators available.

CHAPTER 27: ADDITIONAL FACTS

In case you want to monitor your generation data, there is provision for that in your inverter.

All you need to do is provide an internet connection to your inverter via ethernet or wi-fi.

Some inverter brands provide this service for free and some brands charge a fee.

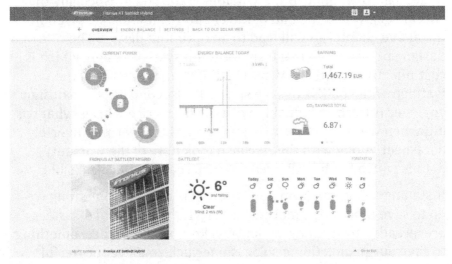

You will need to register and login on the respective brand website to use this service.

CHAPTER 28:
ARE WE DONE?

Learning solar design is a complex task. It's more a mixture of art and science than science alone. As complexity increases, so does the time and effort. Throughout the book, all we did was explain how a typical solar design engineer works. What I have written in this book is no different to what we do in our daily jobs.

Since this is a complex topic that takes time and practice, it is easier said than done. But if you follow the steps, especially the checklists given at the end of the chapters and the one given in the above chapter; it will lead to an optimum design. The role of a design engineer is to make an optimum system design, which is cost-effective and that's what we did when we changed the wattage of the panels to save space. The second duty of a design engineer is to choose quality with the best price. This is what we did when we studied each parameter in the panel and inverter datasheet. Along with this, we also took note of the warranty details which further allowed us to increase our project's life.

Installing solar panels on your rooftop is not just something we do to save money. It's more than that. Recent natural disasters are already an alarming signal, a wake-up call, to start something to save our planet. Going solar is one such small yet powerful step in saving our planet. When a solar plant successfully generates power for the first time, the sound of the system taking its first breath as it swirls into action is an awesome feeling that we design engineers get when a project we designed on paper is actually implemented!

We all leave a mark on this planet. May it be planting a tree or installing a solar panel. This book was a small effort from our side.

ABOUT THE AUTHORS

Paul Holmes is a solar energy system engineer with over a decade of experience in the solar industry. He has worked for residential and commercial customers all over the globe. He focuses on home energy audits and the design of solar power systems for homeowners.

Shalve Mohile is a solar PV design engineer with an experience of over 7 years in designing and building solar projects, with various system configurations including on-grid, off-grid, and micro-grid systems. Shalve has previously worked as a design engineer with companies such as Tata Power and the Ravin Group. Currently, as a co-founder of Curiouskid Engineering Pvt Ltd, he works with clients across the globe.

Shantanu Mohile is an Electrical Engineer. He is also the co-founder of CuriousKid Engineering Pvt Ltd along with Shalve Mohile. He mainly works on spreadsheets calculating savings due to solar, payback, and financial analysis of the same. Apart from Solar, his other interests lie in computer programming.

Made in the USA
Monee, IL
28 October 2024

68844037R00100